Seabed Mineral Resources and the Economic Interests of Developing Countries

Seabed Mineral Resources and the Economic Interests of Developing Countries

Danny M. Leipziger
James L. Mudge

Ballinger Publishing Company • Cambridge, Massachusetts
A Subsidiary of J.B. Lippincott Company

This book is printed on recycled paper.

International Standard Book Number: 0-88410-049-9

Library Congress Catalog Card Number: 76-19076

Printed in the United States of America

Library of Congress Cataloging in Publication Data

Leipziger, Danny M
Seabed mineral resources and the economic interests of developing countries.

Bibliography: p.
Includes index.
1. Underdeveloped areas—Mineral industries.
2. Underdeveloped areas—Economic policy.
3. Mineral resources in submerged lands.
I. Mudge, James L., joint author. II. Title.
HD9506.A2L38 333.8'2'091724 76-19076
ISBN 0-88410-049-9

Contents

List of Figures

List of Tables

Foreword

Having begun this project as neophytes in the area of seabed resources—and hopefully emerging somewhat wiser—we are indebted to a number of individuals who rendered our task easier. We would like to thank Arnold Rothstein, Sherwood Freezon, John Foster, Jacob Meerman, Constantine Michalopoulos, and Bension Varon for reviewing and commenting on early drafts of the material. Over the past year and a half, we have also benefited from discussions with our colleagues; in particular, we would like to acknowledge the council of Robert Muscat and the assistance of Gregory Fager, Wendy Garling, Maureen Lewis, and Susan Mudge. Thanks are also due Nina Cornell, Stephen Salant, and Steven Vanderveen.

Of course, all the opinions expressed in the book are our own and responsibility for any errors rests with us as well. None of the views expressed necessarily reflect those of the U.S. Agency for International Development or the U.S. government.

Danny M. Leipziger
James L. Mudge

Washington, D.C.
April 30, 1976

Introduction

BACKGROUND

Throughout history, the oceans have been of paramount importance to nations as repositories of food, as transportation routes, and as elements of their national security. Access to and use of the oceans commanded the attention of scholars who centuries ago stressed the basic principle that "the sea is common to all, both as to ownership and as to use. It is owned by no one; it is incapable of appropriation, just as is the air. And its use is open freely to all men" [1]. The principle of commonality of the oceans and their resources has served as the foundation of mankind's approach to ocean and ocean resource issues virtually since the origin of recorded history. Accommodation among nations on ocean issues was a logical corollary and means of implementing the commonality principle.

Up to the twentieth century, the commonality principle did not conflict with the practice of accommodation because, if for no other reason, ocean resources were abundant. But in the last half of the twentieth century, ocean space has begun to take on some of the same significant characteristics of land space. It is increasingly subject to pollution hazards, and its food stocks are limited and threatened. Furthermore, the oceans now have the potential for becoming a significant source of hard minerals and hydrocarbons. Indeed, seabed resources comprise the single greatest store of unclaimed mineral wealth in the world. The time of abundant ocean space and abundant living resources but inaccessible mineral and energy resources is rapidly approaching a new era when man's need for resources and his technological capability to acquire them will, for the first time in history, thoroughly test the commonality principle and the centuries-old practice of international accommodation.

The heightened world focus on the ocean's mineral resources is the result of two distinct trends: the steadily increasing demand for mineral resources and petroleum-based energy sources, and the advancement of technology to the point where huge new and unclaimed resources are now within the reach of modern industry. These new resources lie beneath the oceans, extending seaward from the continental shores, where hydrocarbon deposits—oil and natural gas—are trapped in the subterranean strata, to mid-ocean basins, where manganese nodules lie in vast collections on the surface of the ocean floor like rocks in a stream bed [2]. Manganese nodules contain nickel, copper, manganese, cobalt, and several trace elements of other metals. The oceans contain a number of minerals, but hydrocarbons and manganese nodules are, by an appreciable margin, the two most economically significant categories of minerals [3].

The actual quantity of hydrocarbons beneath the ocean floor will never be known. Estimates have been made, however, indicating how much there is, how much will ultimately be discovered and how much of that which is discovered will be lifted to the surface. It has been calculated that the subsurface of the ocean floor may ultimately yield approximately 2272 billion barrels of petroleum and petroleum equivalents which, at a consumption rate of fifty-six million barrels per day, is equivalent to a 111-year supply of petroleum [4]. The total gross value of these hydrocarbon resources could exceed $20 trillion based on 1975 prices.

Manganese nodules, a new and important deep seabed mineral resource, are small spherical objects, concentrations of which litter huge tracts of the deep ocean floor. Nodules derive part of their economic significance from the fact that they are ubiquitous in many deep ocean-bed areas and lie uncovered on the ocean floor. By 1980, deep-seabed mining should be technically feasible to allow commercial-scale mining to take place in areas of the Pacific with water depths as great as 20,000 feet. Mining in one 1000-square-mile area of the Pacific, for example, could provide a million tons of nodules annually for twenty-two years. This amount could supply approximately 271,000 tons of copper; 325,000 tons of nickel; 56,250 tons of cobalt; and 5.6 million tons of manganese ore [5]. There are said to be hundreds of prime mineral sites and the Pacific Ocean alone is estimated to contain 1.6 trillion tons of nodules.

Manganese nodules tend to be located in deep ocean areas outside the limits of coastal state jurisdiction. In order to gain access to offshore hydrocarbon deposits and manganese nodules, nations have begun to act to secure resource rights, either unilaterally or multilaterally. The multilateral approach has centered on the United Nations' Conference on the Law of the Sea (LOS), the most recent event in a long series of historical milestones.

The *Digest of Justinian*, already noted, is the earliest recorded text on the legal theory of the sea. Slightly more than a millennium after the *Digest*, Hugo Grotius defended the theory of freedom of the high seas in his *Mare Liberum.* Significantly, he exempted, *inter alia*, the territorial sea, *Proximum Mare*, from this position and declared that territorial limits are dictated by the areas that a state can defend from land [6].

The body of legal thought on Law of the Sea matters as set out by Grotius survived essentially intact until the onset of the twentieth century. Starting in 1945, major modifications began to be made in the centuries-old tradition of freedom of the seas. These modifications are summarized as follows [7]:

1945: The United States, in the Truman Proclamation, extended United States jurisdiction over natural resources to its continental shelf "beneath the high seas" (i.e., beyond United States territorial limits). This represented a land acquisition of 700,000 square miles. Other nations quickly followed suit with similar claims.

1958: The first conference on the Law of the Sea was convened by the United Nations. The result was four multilateral conventions, which have since been ratified by most nations. The conventions were on: (1) fishing and conservation of the living resources of the sea; (2) the high seas, which defined freedom of the seas in terms of freedom to navigate through, fly over, fish, lay cables but did not state where national waters ended and the high seas (international waters) began; (3) the territorial seas and contiguous zone, which outlined the rules by which the coastal state may exercise jurisdiction within its territorial sea, but did not specify how far out this was to extend; and (4) the continental shelf, which endorsed the Truman Proclamation by giving coastal states jurisdiction over the resources of their continental shelfs to a depth of 200 meters (656 feet) or beyond that to where the depth of the water permitted exploitation of natural resources.

1960: The Second Law of the Sea Conference did not produce any significant resolutions to ambiguities contained in the 1958 conventions.

1967: The U.N. General Assembly established a Committee on the Peaceful Uses of the Seabed and formed the preparatory committee for the 1974 conference.

1970: In a "Declaration of Principles," the United Nations General Assembly declared that there is an area beyond national jurisdiction not subject to national appropriation, to be governed by an international regime, and to be used for peaceful purposes only. It must be open to all countries, including the land-locked, for use and research. This unanimously passed resolution of 17 December 1970 established the principle of "common heritage"; namely, that the resources of the seabed beyond the limits of national jurisdiction were to be considered the "common heritage of mankind."

1971: The United Nations endorsed a treaty banning weapons of mass destruction from the ocean floor.

1972: The General Assembly called for a comprehensive conference in 1974 to deal with all aspects of sea law, legal as well as economic.

1974–1976: The Third United Nations Law of the Sea Conference met in sessions in Caracas, Geneva and New York.

The most important point in the evolution of the Law of the Sea with respect to mineral resources is the 1970 "Declaration of Principles," which referred to resources of the sea in the area beyond national jurisdiction as the "common heritage of mankind" [8]. There are two issues in this declaration which are crucial to an inquiry into the economics of ocean hydrocarbons and manganese nodules. First, what is the operational interpretation of the "common heritage of mankind?" and second, what is the seaward limit of national jurisdiction? The interpretation of the common heritage concept has powerful implications for the potential allocation of gains between rich and poor nations from exploitation of seabed hydrocarbons and manganese nodules. The seaward extent of national jurisdiction and the interpretation of that jurisdiction will determine what portion, if any, of seabed hydrocarbons will be included within the common heritage realm and thereby will determine the basic distribution of gains between a few coastal states and the international community.

There can be little doubt that the Law of the Sea negotiation is one of the most important conferences since the formation of the United Nations; certainly, it is viewed with tremendous import by the Third World. Because the Third LOS Conference deals specifically with allocation of the huge stock of natural resources in the international area, the conference provides a unique opportunity for international economic cooperation between developed and developing countries. By present indications, however, not only are the developed and developing countries far apart on some issues, but, significantly, the richer and more influential countries—both developed and developing—are pursuing policies which might result in a smaller total share of resources for much of the developing world. In particular, a group of the poorest economies of the world, many of which are geographically disadvantaged, appear likely to gain very little in the allocation process.

An overview of the positions taken by LOS participants reveals a somewhat peculiar situation. There are three interrelated elements of interest. First, the approximate magnitude of the gross value of manganese nodules and hydrocarbons are $3 trillion and $20 trillion, respectively. Second, because the difference between the selling price of petroleum and its production cost tends to be much greater than the difference for manganese nodules, the net revenue potential of hydrocarbons is much greater than a comparison of gross values might suggest. In spite of the vastly greater wealth potential of hydrocarbons, control over production of manganese nodules and sharing in the benefits of manganese nodule production have dominated the debate on exploitation of mineral resources in the LOS negotiations.

A second point is that the LOS negotiations have not consistently followed the standard pattern of developing country–industrial country debate. The line of debate on manganese nodules follows the now routine alignment in which developing countries are arrayed against the industrial economies. The international alignment on division of hydrocarbon resources differs radically, however, and it tends to be dominated by geographical factors as opposed to economic factors, with coastal states, regardless of economic status, relatively unified in their aims to acquire economic control over their adjacent coastal seabed areas.

The last point is that the length and intensity of debate between coastal and non-coastal states (land-locked, shelf-locked, and zone-locked) over control of seabed hydrocarbon resources has been limited and muted. Land-locked and geographically disadvantaged states, which include economies at all stages of development, have gradually increased the level of their participation in the debate, however, as sessions of the LOS Conference have progressed from Caracas to Geneva to New York. The disadvantaged group has increasingly concentrated its attention on the question of acquiring a portion of the gains from production of mineral resources, while accepting in principle the creation of economic jurisdiction zones by the coastal states [9]. Nevertheless, the mainstream of debate in these sessions has been directed toward the role of developing countries in manganese nodule exploitation. This is not to say that exploitation of manganese nodules is not a significant issue and does not deserve the attention of developing economies, but given the focus of the debate, the relative value of the two types of resources, and the international alignment of nations, it is not unrealistic to expect that the final outcome will see landlocked and geographically disadvantaged states receiving a small share of the smaller resource-manganese nodules. On the other hand, the coastal nations, including many of the industrial and virtually all of the resource-rich developing countries, may be able to divide all of the major resources among themselves and still obtain a substantial share of the minor resource.

APPROACH

In recent years, the literature on seabed mineral resources has largely focused on legal, political, and technical aspects. The few economic treatments have tended to be relatively restricted, concentrating on one or two specific areas such as revenue-generating potential or revenue distribution schemes. One of the main purposes of this undertaking is to make a first attempt at providing a reasonably comprehensive and systematic examination of the major economic issues related to allocation and production of the virtually untapped stock of seabed hydrocarbons and manganese nodules.

This study also pursues another objective, which is to attempt to contribute to the body of economic information for use by countries in their international negotiations on seabed resource issues. Specifically, in a negotiation as wide ranging and complex as the Law of the Sea, in which limits of coastal state jurisdiction over offshore hydrocarbon resources and international control over deep seabed resources will be determined, it is crucial that countries bargain with the best possible information at their disposal. Determination of political best-interest is at times easier than the determination of economic best-interest because the information and analytic underpining necessary for economic evaluation are by comparison more scarce than the availability of material necessary to make a political decision. In the case of seabed mineral resources, economic information is scanty. For many countries, the lack of economic information allows political considerations to dominante economic interest, potentially to the detriment of their long-term economic interest. Each country participating in the LOS negotiations has a unique set of economic and political factors influencing its position and each country must evaluate proposals with respect to its particular set of political-economic policy criteria. Because of the complexity of the negotiation process and the propensity for economic considerations to be less weighty than political aspects in the determination of the country's position, an examination which concentrates on the economic aspects of seabed minerals should prove helpful for many countries still weighing their own policy trade-offs and evaluating the proposals put forward in the course of international discussions on economic jurisdiction over seabed resources.

It is also essential to explain the particular nature of the economic approach employed here. The material in the preceding background section clearly indicates that the recent Law of the Sea negotiations represent an important milestone (but hopefully not a capstone) in the economic evolution of seabed mineral resources and the outcome of the negotiations themselves are likely to exert a strong influence on the economic resolution of seabed resource in the twentieth century. Thus, examination of the economic issues throughout the book will use the negotiations as the backdrop for discussion. The examination, however, is not intended as a study of the institutional aspects of seabed mineral resources. Rather, it is intended essentially to apply basic economic principles in a critical evaluation of the exploitation of natural resources in an international area. "Political realities" are relegated to a secondary position in order to facilitate a more complete economic evaluation of the issues.

The underlying framework of reference in the course of our analysis is efficiency and equity. Efficiency refers to how resources are allocated in production, while equity refers to how goods are distributed to consumers. If the world economy could be characterized as a perfect or nearly perfect competitive system, a distinction between efficiency and equity would not be necessary; producers and consumers, acting in their self-interest, would, through markets for factors of production and for goods, allocate global resources in an optimal fashion. No reallocation of resources could improve the final outcome for society.

Because the world economy in many major aspects diverges from a competitive ideal, we have chosen to stress production efficiency in the traditional positive sense and to combine this with a normative approach to equity. Production is considered efficient if firms are free to select output levels which maximize profits; such an output decision is potentially an optimal allocation of resources. Maximum use of competitive markets is still assumed to be the approach which best approximates an efficient production solution.

The most serious impediment to efficient distribution of output is the pervasive disparity among income levels worldwide. This is especially true if the output being produced—seabed minerals—is predominantly consumed by the high income nations, yet is considered the "common heritage" of mankind. Where traditional methods of distribution fail, equity concerns must be directly addressed; for example, through international taxation and redistribution of revenue. Rather than interfere with efficient production, it is possible to redistribute gains in a way which responds directly to equity concerns. In this manner, we hope to recognize the salient political and institutional factors while simultaneously pursuing a systematic economic inquiry.

FRAMEWORK

The first half of the examination focuses on hydrocarbons beneath the ocean floor. Hydrocarbon exploitation has not been a distinct issue in the Law of the Sea, but rather has been submerged in the debate over the creation of economic zones which would extend seaward from each coastal state's territorial sea to encompass an area which would include all or part of the submerged portion of the continent. Each country with a coast would acquire the exclusive right to control exploitation of the resources in the adjacent economic zone, including those hydrocarbons which might exist in the substrata.

The creation of economic zones immediately creates two subgroups of developing nations: (1) countries which have no offshore hydrocarbon prospects whatsoever because for geographical reasons they would not acquire economic zones, and countries with economic zones which for geological reasons are devoid of commercial hydrocarbon potential; and (2) a second subgroup consisting of coastal developing nations with the potential for commercial offshore petroleum development. The economic interests of the first disadvantaged group, consisting of approximately fifty-six countries, would appear to support internationalization of a substantial portion of hydrocarbons in the continental margin or at least favor international sharing of the benefits from offshore hydrocarbon production in areas relinquished by the international community. The more parochial interests of the second group of states are virtually identical to those of the industrial countries with similar geographical and geological advantages; each country may be viewed as desiring to maximize the area of its economic zone (expressed in practice as the seaward extent of the zone) in order to maximize the potential resource gain and, consistent with this gain, each coastal state would prefer to minimize any sharing of benefits from the production of hydrocarbons from its economic zone. This fundamental but virtually overlooked dichotomy between petroleum "haves" and "have nots" of the Third World suggests a number of avenues for economic inquiry.

The analysis commences in Chapter 1 with an examination of the distribution of undersea hydrocarbon reserves with particular emphasis on the distribution of offshore hydrocarbon potential relative to per capita income. Many coastal states of the Third World (as well as many industrialized countries) are adamant on extending the seaward extent of their economic zones, but what are the prospects of gaining access to hydrocarbon resources as opposed to merely acquiring underwater real estate devoid of hydrocarbon potential? If the seaward extention of exclusive national resource claims is eventually institutionalized, some states may find that, at least as far as hydrocarbons are concerned, they would have been better off to stress internationalization of seabed resources (with revenue sharing) instead of nationalization. The distribution of offshore petroleum potential in relation to national income per capita is of particular interest from the point of view of economic development. Examination of offshore hydrocarbon potential relative to national per capita income provides a response to the question of whether offshore petroleum production offers at least a partial offset to national income differences or whether the distribution of offshore potential would merely tend to exacerbate the prevailing differences in income distribution among developing nations. A companion issue is whether the distribution of offshore potential throughout the world tends to reinforce or offset the prevailing pattern of income distribution worldwide.

Aside from the geographical and geological facts which affect low-income country prospects, there are also the challenges of technical capability and commercial feasibility, the topic of Chapter 2. The issue of technical capability has important implications for developing economies both in terms of what the technical constraints are on the scope of exploitation and what the prospects are for participation in offshore hydrocarbon exploitation. The issue of technical limits has implications for disadvantaged states as well as for those nations with promising offshore areas. Technical capabilities are important to the geographically disadvantaged countries because, if the latter group are excluded from near or intermediate offshore areas, their only hope is that hydrocarbon exploitation may be feasible in the more distant and deep offshore areas.

There are four major links in the chain of events leading to offshore hydrocarbon production: (1) possession of rights to extract the resources of a section of continental margin; (2) the discovery of hydrocarbon-bearing substrata in the margin; (3) technical capabilities to exploit the resources; and (4) the financial viability of exploitation. Commercial feasibility is the most crucial issue of deep offshore petroleum extraction, given the existence of a strong resource potential. There is a relatively broad consensus that technical capabilities are substantial, whereas there are differing opinions on the commercial feasibility [10]. The viability link is in a sense more complex because its disposition is a function of several variables, most of which can be expected to change over time. In spite of the uncertainty and the debate over commercial feasibility, so far, no methodical effort has been made to compare deep offshore exploration and development costs to petroleum prices in an effort to gain some insight into the commercial feasibility aspect. Examination of the commercial feasibility question is particularly interesting in view of the major increases in petroleum prices since 1973 and 1974.

Chapter 3 focuses on the degree to which developing economies can be expected to participate in offshore activities, especially in deep water offshore petroleum production. The initiation of large-scale offshore operations coincides with the maturation of some major, conflicting trends. On one hand, developing nations have shown a greatly increased interest in managing and implementing their own national petroleum policies. On the other hand, deep-water offshore operations require advanced technologies and very large amounts of investment both of which tend to be associated with industrial country sources. These conflicting trends are examined with respect to a spectrum of possible petroleum development policies, as well as with respect to the financial and technical capacities of the nations of the Third World.

While the degree of developing economy participation in offshore operations is important, a significant overriding consideration is the probable economic impact. While hydrocarbon production is a seemingly unmitigated benefit to a country, some factors may be less beneficial than others. Also there is a question of what economic effects might result for those low-income countries which do not develop offshore production.

Chapter 4 is an inquiry into the debate over allocation of the rights to seabed resources and an examination of some of the economic implications flowing from the debate. After some early initiatives to reserve a large portion for international community purposes, the debate appears to have evolved into a stampede to divide these resources essentially along national lines. The move toward nationalization of seabed resources is extremely relevant in the case of hydrocarbons because they are predominantly located in the continental margin and would thus be encompassed by nationally controlled economic zones. Recognizing that a significant proportion of subsea hydrocarbon reserves will probably be nationalized, it becomes important to examine the ramifications of nationalization in terms of the economic interests of low-income countries which are excluded from the opportunity to exploit offshore hydrocarbon resources.

It has been suggested that some of the revenues from seabed mineral production might be shared with the international community, especially with the less developed economies. But there has been little effort to highlight either the unique advantage acquired by states which gain exclusive rights to offshore hydrocarbons or, in the absence of any sharing mechanism, how international income disparities might be affected. Furthermore, little explicit attention has been given to consideration of what claim, if any, geographically disadvantaged states have to resources found within the continental margins. If international revenue sharing were adopted as a major feature of seabed hydrocarbon production, a number of complex economic issues arise relating to the revenue collection and distribution systems. In view of the seemingly huge value of subsea hydrocarbon resources, it is also important to examine the potential revenue-generating capacity of these resources.

Deep seabed mineral resources are unique in a number of important respects and the following three chapters are devoted to those issues. First, the deposits of manganese nodules are predominant in areas not claimed by coastal states and have thus been characterized as "common resources" for mankind. Second, the experimental and highly capital-intensive technology necessary to exploit nodules commercially is being developed by a few industrialized nations, so that the gains of manganese nodule exploitation are likely to flow primarily to a few developed nations. Third, the other possible beneficiaries of nodule production are the major consumers of the minerals contained in nodules—nickel, copper, cobalt, and manganese—which are predominantly the industrialized countries. And fourth, the only potential losers from nodule production are the land-based mineral producers, which are in most cases developing nations. Thus, developing nations stand to gain little—and in some cases even to lose—from the exploitation of deep seabed minerals.

An understanding of these characteristics of nodules is necessary for any analysis of deep-seabed mining, which, it is anticipated, might begin at the end of this decade. A handful of mining firms have invested significant sums of money to develop deep ocean mining systems, capable of operating in water depths up to 4 miles below the ocean surface. These firms are, of course, vitally concerned that a stable deep-seabed regime be established to guarantee the security of these investments, estimated to be approximately $1 billion by 1980. The industrialized nations, likely seabed mineral producers and major minerals consumers, are thus interested in securing free access to deep ocean resources in an atmosphere free of international constraints. Thus, their primary concern is for production efficiency and unimpeded access to the common resources.

Developing nations are vitally concerned that they participate in the gains to be derived from deep-seabed mining. Inasmuch as they are unlikely to benefit from the direct production of manganese nodules by developed nations, they are less concerned with production efficiency than with the distribution of gains of seabed mining. Thus, the developing nations propose an International Seabed Resource Authority (ISRA) which would control all aspects of deep-seabed mining and would thus collect much of the gains of seabed mining. In attempting to garner the benefits of nodule mining for all nations, the developing nations, as represented by the Group of Seventy-Seven (their negotiating unit in the U.N. Conference), are willing to trade off efficiency of production.

In order to analyze deep-seabed issues, one needs to explore the basic facts concerning the nature and location of manganese nodules and concerning the development of a deep-seabed mining technology. In addition, there is the important issue of developing country dependence on land-based mineral exports and the potential for adverse economic consequences as a result of nodule mining. These issues are explored in Chapter 5.

One of the major difficulties in attempting to negotiate an international deep-seabed mining regime is that it is difficult to forecast the nature of manganese nodule mining, which has yet to begin. Commercial production is not anticipated to reach significant levels until the mid-1980s, and therefore it is hard to estimate the profitability of mining or the magnitude of the potentially adverse impact of seabed mining on land-based mineral producers. These two factors are of predominant importance in constructing a deep-seabed mining system. The issue of adverse impacts of seabed mining is of paramount interest to major copper, cobalt, and manganese exporting developing countries. In Chapter 6, price and revenue projections are developed for nodule mining in 1985 and are compared to estimated costs to derive profitability estimates. The possible impact of manganese nodule mining on land-based mineral producers is also examined, and measures are reviewed which might assist developing country mineral producers.

One method of structuring a deep-seabed mining system to allow efficient resource exploitation, which is one objective of an international resource management system, while still protecting the interests of developing countries (especially mineral producers), is to construct a system of international taxation. An international deep-seabed mining tax would redistribute the gains of production to those nations not normally sharing in the benefits but still entitled to a portion of the gains from the "common heritage of mankind." A portion of such taxation, not used for revenue sharing, would be allocated to a compensatory facility to assist developing country mineral producers to diversify their export profiles. This is one possible compromise solution between a potential seabed monopoly and a system of free access (without any sharing of benefits), one which attempts to combine efficiency of production with an equitable distribution of gains. Chapter 7 explores the positions of the developed and developing countries on the characteristics of a deep-seabed mining regime within the context of the Law of the Sea Conference. We attempt to provide a compromise position, based on both efficiency and equity concerns, which addresses the vital interests of both groups of countries within a multilateral framework. In so doing, an analysis is provided of alternative international taxation mechanisms. Projections of potential tax revenue and possible uses for these funds are also examined in Chapter 7.

Chapter 8 commences with a brief examination of the economic interrelationship between hydrocarbon and manganese nodule resources and then offers some reflections on the 1976 Law of the Sea negotiations and the extent to which efficiency and equity concerns are being addressed in this multilateral forum. Major trends in the negotiations over rights to hydrocarbon and deep-seabed mineral resources are examined with respect to the implications for developing nations.

NOTES TO INTRODUCTION

1. The Digest of Justinian (529 A.D.) as reported in Percy T. Fenn, *The Origin of the Right of Fishing in Territorial Waters*, Cambridge: 1926, p. 23.

2. The Arctic and Antarctic areas are excluded from both the discussion of hydrocarbon resources and the discussion of manganese nodules.

3. The oceans, the oceans' floors, and the substrata of the seabed contain many of the minerals found on land, including salt, magnesium, and uranium in salt water; diamonds, tin, sulphur, iron sands, and phosphate on or near the ocean floor; and coal in the substrata of the seabed. A.A. Archer, "Economics of Off-Shore Exploration and Production of Solid Minerals on the Continental Shelf," *Ocean Management* 1 (December 1973): 5–40. The potential value of these resources in particular locations could be substantial, but their aggregate value and international economic significance does not compare to that of hydrocarbons or manganese nodules.

4. The 2272 billion barrel estimate was made by Lewis G. Weeks in 1973. United Nations (See p. 20, note 11).

5. J. Mero, "The Mineral Resources of the Sea" (Elsevier, 1964), p. 175; M. Dubs, J. Flipse and R.J. Greenwald, "Pre-Production Manganese Nodule Mining Activities and Requirements," in *Mineral Resources in the Deep Seabed*, Hearings of the 93rd Congress: 1 (May-June, 1973), pp. 602–709; and A. Rothstein and R. Kaufman, "The Approaching Maturity of Deep Ocean Mining—The Pace Quickens in *Mineral Resources in the Deep Seabed, Supra.*, pp. 201–222.

6. P. Fenn, *The Origin of the Right of Fishing*, p. 165.

7. Adapted and quoted from: United Nations, Center for Economic and Social Information, *Development Forum*, Volume II, May 1974, p. 2.

8. United Nations, General Assembly, *Declaration of Principles Governing the Sea-Bed and the Ocean Floor, and the Subsoil Thereof Beyond the Limits of National Jurisdiction* (17 December 1970).

9. Austria, acting as spokesman for the land-locked and geographically disadvantaged countries, proposed that there be revenue sharing from exploitation of non-living resources of the continental shelf. Austria also proposed a specific distribution scheme for shared revenues from manganese nodule production.

10. John King Gamble, Jr., and Giulio Pontecorvo, eds., *Law of the Sea: The Emerging Regime of the Oceans* (Cambridge, Mass.: Ballinger Publishing Company, 1973), pp. 6, 13.

Chapter 1

The Worldwide Distribution of Hydrocarbons

The particular locational pattern of hydrocarbon resources lying within the underwater portions of the continents exerts a fundamental influence on the Law of the Sea deliberations.

If offshore hydrocarbons were evenly distributed, or if they were located exclusively in offshore areas which traditionally have been the economic domain of coastal states, the hydrocarbon issue would be of minor significance in the LOS debate. But offshore hydrocarbons are not distributed evenly among nations, and a substantial portion of these resources lie in areas which are now, and have been throughout history, in an international area. The distribution issue is further complicated by large differences in relative wealth and the relative technical capacities among the nations of the world. If all nations were on a relatively common plane in their ability to exploit and their need for hydrocarbon resources, the issue would lose a degree of complexity, even if the location pattern were not symmetrical.

The purpose of this chapter is to examine several different aspects of offshore hydrocarbon distribution—including distribution relative to national income. Before the distribution discussion can be taken up, however, it is necessary to note some relevant geological facts and to delineate some terms used in the discussion of hydrocarbons.

ASPECTS OF THE GEOLOGY OF THE OCEAN FLOOR

About 70 percent of the earth's surface is seabed and about 21 percent of the seabed consists of an area known as the continental margin [1]. The continental margin is that portion of the continents which extends below the surface of the ocean and is the part of the ocean floor most likely to yield large amounts of hydrocarbons. As in the continental masses above sea level, there are sedimentary formations—the type of geological formation in which hydrocarbons are found—beneath the margin surface. The key geomorphic features of the continental margin are depicted in schematic form in Figure 1–1.

The actual profile of the world's continental margin varies widely; in some places the shelf and the slope may not be clearly differentiated. For the purpose of considering hydrocarbon-bearing potential, the thickness of sedimentary formation in the margin area is probably the single most important factor. The sediment layer may

Figure 1–1. Geomorphic Profile of the Continental Margin

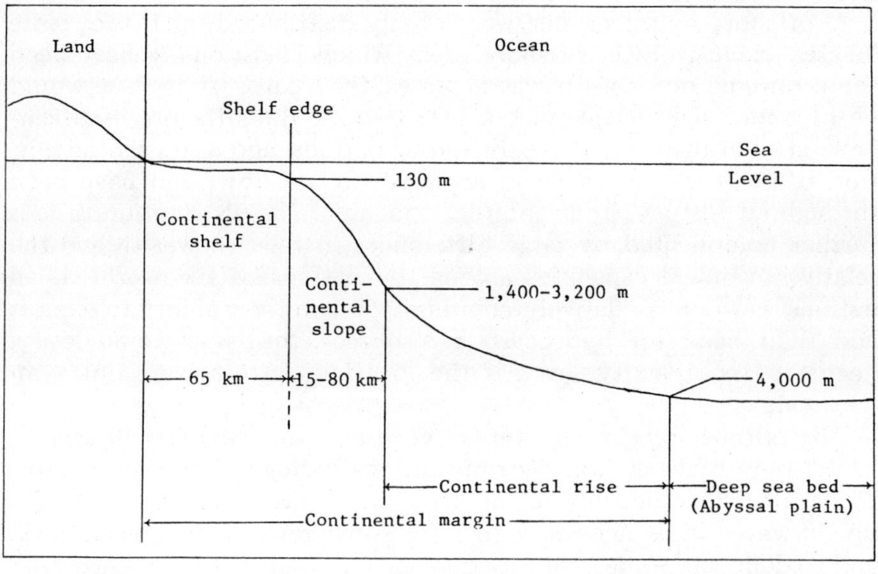

Source: Adapted from U.S. Department of Interior, United States Geological Survey, *The Worldwide Search for Petroleum Offshore: A Status Report for the Quarter Century 1947–72*, by Henry L. Berryhill, Jr., Geological Survey Circular 694 (Washington, D.C.: Government Printing Office, 1974), p. 5.

be as thick as ten kilometers and although there is a great deal of variation, sediment layer thickness frequently is found to increase in the outer shelf and slope regions becoming thinner as the continental rise approaches the abyssal plain [2]. It is significant in the context of the Law of the Sea and consideration of the seaward extent of national jurisdiction that sedimentary volumes beneath rises and slopes are likely to be substantially greater than sedimentary volumes beneath the shelves of the continental margins. The difference in sedimentary formation thickness implies that the hydrocarbon potential of the continental slope and rise may be greater than that of the continental shelf, where virtually all previous offshore exploitation activity has taken place. The possibility that the continental margin beyond the continental shelf may have appreciable potential for hydrocarbon production suggests that non-coastal states ought to have an active interest in the efforts of coastal states to extend their economic rights seaward into the international area where the continental slope and rise lie.

In any discussion of petroleum resources, there are a number of terms which have specific meanings indicating the categories of hydrocarbon resources. These terms delineate, for example, whether the actual location of the resource has even been discovered and whether it is extractable using current techniques at current costs and prices. A set of terms which describes the entire geological-economic spectrum of hydrocarbons lends perspective to the one or two terms which will constitute the primary focus of the discussion which follows:

Total Resources. Hydrocarbon deposits which have not yet been discovered (including that portion which will never be discovered), deposits which have been discovered but the exploitation of which is not financially viable, and proved reserves.

Ultimately Discoverable Resources. That portion of resources which is estimated eventually to be discovered.

Undiscoverable Resources. Hydrocarbon deposits, the size and location of which will remain unknown.

Potential Ultimate Production. The amount of hydrocarbon which will in fact be produced.

Unrecoverable Resources. That portion of known hydrocarbons which will be left underground.

Potential Production. That portion of ultimately discoverable resources which can be recovered in the future. This category of hydrocarbons will be the main focus of discussion [3].

As an example of how the various categories might relate to one another, it has been estimated that for the United States offshore area, extending to a depth of 500 meters, approximately 62 percent of all hydrocarbon resources will be ultimately discovered. The remaining 38 percent will remain undiscovered. About 43 percent of ultimately discoverable resources are expected to be recoverable (27 percent of total resources), and the remaining 57 percent (36 percent of total resources) will be left in the ground [4].

As a final point, it is noted that application of the term "offshore" is frequently based on a superficial locational aspect of production, i.e., if the production site is not located on dry land, then it is "offshore"—even if the offshore site is a shallow tidal marsh or an inland lake. The determination of offshore on this basis also means that offshore wells extracting petroleum from an onshore reservoir which extends seaward from the coast are also considered as offshore production. The United States Geological Survey (USGS) offers a narrower definition of offshore in which underwater hydrocarbon deposits that are obviously within the national domain of a nation are not considered to be offshore. The USGS defines offshore hydrocarbon-producing areas as:

> 1) waters along shores facing high seas, or 2) waters along shores of relatively confined seas through which an international boundary exists. Bodies of water surrounded by one nation or inland seas are not regarded as offshore waters. Following this definition, wells drilled or producing in Lake Maracaibo in Venezuela or the Caspian Sea in the Soviet Union are assigned to the onshore classification [5].

Attention is drawn to the more rigorous definition of offshore because most offshore production and reserve data (except for the USGS data) are based on the broader interpretation of offshore. The USGS definition, however, is a more appropriate concept for use in an international context and therefore is adopted for the purpose of this discussion.

THE RECORD OF OFFSHORE HYDROCARBON PRODUCTION AND EXPANSION

World offshore oil output in 1972 totaled about 2.4 billion (10^9) barrels, and natural gas production totaled 24,000 billion cubic feet.

Offshore oil production accounted for 13 percent of world total production in 1972 [6]. Approximately 5600 billion cubic feet of natural gas, or about 12 percent of total output, was produced from the offshore area in 1972 [7].

In 1972 the ratio of cumulative onshore oil production to cumulative offshore oil production was 20.5:1. The onshore-offshore ratio of proved recoverable reserves, however, was 4.5:1. Thus the relatively small offshore area which had been tapped by 1972 was extremely important in terms of reserves. Table 1–1 indicates the regional distribution of offshore oil production and reserves in 1972.

The extraction of hydrocarbons from beneath the seabed is a relatively recent activity. Onshore oil production started in the 1870s, but, excluding shallow-water operations which tap onshore deposits, the first offshore oil reservoir started producing only in 1947 in the Gulf of Mexico. In 1965 only five countries had recorded offshore oil or gas production. By 1969, however, more than seventy-five nations were conducting exploratory work in their coastal areas and twenty-two were producing hydrocarbons from offshore sites [8]. In 1974 oil was being produced off the coasts of thirty-one nations, and offshore exploration was being conducted in the waters of more than sixty countries [9].

Starting with a single well in 1946, the growth in offshore hydrocarbon production has been very rapid, as the data in Table 1–2 show. During the decade of the 1960s, the growth in offshore hydrocarbon production mushroomed. Offshore oil and natural gas production from the Middle East increased by a factor of over 16, while oil production from United States coastal areas expanded more than threefold and natural gas output was seven times greater in 1972 than it was in 1961. Africa and Asia had little or no offshore production in 1961, but within eleven years these regions had become major producing areas with particularly large output increases between 1971 and 1972.

The rapid growth was accompanied by an increase in the regional dispersion of output. In 1961 the United States was the dominant producer, with 59 percent of total oil output and 85 percent of natural gas production. Over the decade, the regional distribution of production became much broader, and in absolute terms the Middle East became the dominant producer of oil, although the United States maintained a preeminent position in natural gas production. By 1972 the Middle East was producing 56 percent of total offshore oil output, followed by the United States with 20 percent. Also by 1972, three newcomers had appeared and had attained substantial production levels: output of oil from Africa was 13 percent of total

Table 1–1. Worldwide Offshore Production and Recoverable Reserves, 1972

	Production (mil. of 42 gal. bbls.)	*Recoverable Reserves (mil. of 42 gal. bbls.)*	*Recoverable Reserves (percent of World Total)*
Western Hemisphere	*521.4*	*10,917.6*	*11.69*
Mexico	25.8	1,934.0	2.07
United States	472.3	7,664.0	8.20
Brazil	3.5	19.6	.02
Peru	10.1	n.a.	—
Trinidad & Tobago	9.7	1,300.0	1.40
Venezuela	—	n.a.	—
Africa	*309.4*	*4,782.0*	*5.08*
Angola	46.5	n.a.	—
Congo	1.8	730.0	.78
Egypt	58.4	n.a.	—
Gabon	19.9	652.0	.70
Nigeria	182.8	3,400.0	3.60
Europe	*13.4*	*12,250.0*	*13.20*
Denmark	.6	250.0	.30
Italy	.7	n.a.	—
Norway	12.1	7,000.0	7.50
United Kingdom	—	5,000.0	5.40
Asia	*1,565.2*	*65,166.0*	*70.03*
Brunei	45.2	2,413.0	2.60
Indonesia	21.0	400.0	.43
Iran	258.0	5,000.0	5.40
Japan	.4	15.0	—
Kuwait	75.0	n.a.	—
Malaysia	33.6	n.a.	—
Qatar	87.7	n.a.	—
Saudi Arabia	757.0	57,338.0	61.60
UAE	173.3	n.a.	—
Australia	114.0	n.a.	—
Total	*2,409.4*	*95,115.6*	*100.00*

Source: U.S. Department of Interior, United States Geological Survey, *Summary of 1972 Oil and Gas Statistics for Onshore and Offshore Areas of 151 Countries*, by Sherwood E. Frezon, Geological Survey Professional Paper 885 (Washington, D.C., 1974), *passim.*

world output, while the shares of Asia and Europe were 9 percent and 6 percent, respectively.

The output pattern of natural gas in 1972 probably reflects the proximity to consuming markets more than it does the relative production ability of the various regions. Thus the United States and Europe in 1972 were producing 59 percent and 21 percent of total

Table 1–2. Growth in Recorded Offshore Hydrocarbon Production, 1961–1972 *(Oil figures in 1 million barrel (10^6) units; natural gas in billion (10^9) cubic feet)*

	1961		1966		1971		1972	
Region Categories[a]	*Oil*	*Gas*	*Oil*	*Gas*	*Oil*	*Gas*	*Oil*	*Gas*
Middle East	84.0	47.3	320.5	180.3	1,107.9	333.7	1,351.4	808.6
Africa	—	—	2.1	—	129.4	—	309.4	206.2
Europe	—	—	—	10.6	2.0	654.9	13.4	1,178.4
United States and Canada	133.4	478.0	297.7	1,372.0	548.6	3,667.0	472.3	3,325.0
Latin America and Caribbean	9.3	35.2	18.8	42.2	42.4	61.9	49.1	57.6
Asia[b]	.6	3.5	.7	3.5	124.1	35.8	213.8	31.4
Total[c]	227.3	564.0	639.8	1,608.6	1,954.4	4,753.3	2,409.4	5,607.2

Source: U.S. Department of Interior, United States Geological Survey, *Summary of 1972 Oil and Gas Statistics for Onshore and Offshore Areas of 151 Countries*, by Sherwood E. Frezon, Geological Survey Professional Paper 885 (Washington, D.C., 1974), *passim*; U.S. Department of Interior, United States Geological Survey, *Summary Petroleum and Selected Mineral Statistics for 120 Countries, Including Offshore Areas*, by John P. Albers *et al.*, Geological Survey Professional Paper 817 (Washington, D.C., 1973), *passim.*

[a]Region categories are as follows: Middle East (Egypt, Iran, Israel, Kuwait, Qatar, Saudi Arabia, United Arab Emirates); Africa (Angola, Congo, Gabon, Nigeria); Europe (Denmark, Italy, United Kingdom, Norway); Latin America and Caribbean (Mexico, Brazil, Peru, Trinidad & Tobago); Asia (Brunei, Indonesia, Malaysia, Japan, Australia, New Zealand).

[b]Quantities are understated to the extent that some output data were not available. In most of these cases output was probably relatively small.

[c]Data are not available for the Soviet Union and the People's Republic of China. The USSR did indicate offshore petroleum production of 81 million barrels in 1971, whereas the offshore output from the PRC was probably insignificant.

offshore production, respectively, followed by the Middle East with over 14 percent and Africa with 4 percent.

Overall, offshore oil and gas production approximately doubled each year from 1961 to 1972. The traditional producing regions of Latin America and the United States experienced relatively slow rates of output growth, while total output from regions for which offshore exploitation was a relatively new activity (Africa, Europe and Asia) had by 1972 surpassed the United States in total offshore petroleum output and had attained a level of natural gas production equal to one-half of the United States level.

The data in Table 1–3 illustrate the shift in offshore drilling activity (using the broad interpretation) from traditional to new areas (i.e., from the United States and Latin America to Asia and Europe) for a recent six-year period. The overall growth rate remained at a relatively constant level of about two thousand wells per year. In the Western Hemisphere, which includes the two regions accounting for almost three-fourths of total offshore drilling, the drilling rate remained constant. In the Far East, however, offshore drilling activity approximately doubled between 1969 and 1972–73. The major portion of this increase can be attributed to Indonesia, but offshore drilling also accelerated off the coasts of Malaysia, Thailand and Australia. Over the 1969–1973 period, the annual drilling rate also doubled in the North Sea, with Norway and the United Kingdom as the main participants.

WORLD OFFSHORE HYDROCARBON POTENTIAL AND DISTRIBUTION WITH RESPECT TO WATER DEPTH AND DISTANCE FROM SHORE

To obtain some perspective on what the continental margin has to offer, it is useful to consider the amount of hydrocarbons which might be produced from the margin. The amount of hydrocarbons is presented as an estimate of production potential, a concept which incorporates a number of supporting assumptions and subsidiary estimates. The estimate consists of two basic elements: a calculation of how much hydrocarbon exists and an estimate of the quantity which will be produced. The estimation of hydrocarbon production potential has been characterized as "an imprecise and speculative art, fraught with so much subjectivity and intuitive interpretation that the subject has been compared with crystal ball gazing" [10]. While a rather strong but probably not altogether unjustified characterization, the crucial point is that the precision implied by a specific

Table 1–3. Growth in Worldwide Offshore Drilling *(Number of new wells)*

	1968	*1969*	*1970*	*1971*	*1972*	*1973*
Africa	80	85	21	59	111	90
Canada and Latin America	425	438	489	512	784	444
Europe	85	85	55	20	279	171
Far East	171	226	87	103	443	346
Middle East	95	100	62	44	117	53
United States	1,487	1,341	1,063	916	926	1,029
Other	12	15	1	11	1	—
World Total	2,355	2,290	1,778	1,665	2,661	2,133

Sources: "Worldwide Statistics," *Offshore* 30 (20 June 1970):50–57; "Worldwide Statistics," *Offshore* 34 (20 June 1974):88–89.

number indicating production potential should not be taken literally, but should instead be interpreted as an order of magnitude of production potential.

Estimates of the world's offshore hydrocarbon production potential have consistently risen over time. The increase is essentially a reflection of the introduction of new knowledge and technological improvement. The estimate is made by first calculating a yield factor based on experience with similar geological formations. Then the yield factor is applied to volume estimates of subsea sedimentary formations to derive a figure representing hydrocarbon production potential. Estimates of production potential increase as yield factors increase, as more knowledge is generated about sedimentary formation volumes, and as the size of the offshore area under study is expanded (because it becomes possible to exploit hydrocarbons at greater water depths).

In 1965 Lewis G. Weeks estimated that the world offshore area extending to a water depth of 305 meters (1000 feet) would yield a production potential of 700 billion barrels of petroleum liquids [11]. Reflecting the evolving situation, in 1971 Weeks increased this estimate to 1550 billion barrels for the same area of seabed. This estimate incorporated 790 billion barrels of petroleum, an additional 360 billion barrels which were assumed to be recoverable by secondary methods (a technique which frequently involves injection of water or steam into wells to recover oil which otherwise could not be lifted), and 400 billion barrels (equivalent quantity) of natural gas [12].

In 1973 the estimate of the amount of recoverable petroleum was increased to 2272 billion barrels of oil and oil equivalent. The change in part reflects the extension of the offshore area from the 305-meter isobath seaward to the 3000-meter isobath [13]. Weeks' assumption that the entire 2272 billion barrels are economically recoverable is not without an element of optimism in view of the fact that the estimate includes petroleum formations from under as much as 3000 meters of water, and the technique for production of petroleum at depths much greater than 1000 meters has not yet been fully developed. On the other hand, recalling that petroleum prices quadrupled in late 1973—early 1974, Weeks' 1973 estimate of recoverable petroleum may not be overly optimistic.

The distribution of offshore hydrocarbons is important because the seaward extent of the economic domain of coastal states has been a major issue in recent years and one which will presumably be resolved in the course of the LOS deliberations. Distance from shore has traditionally been used as the standard to measure the seaward limit of coastal-state jurisdiction, but in the last two decades ocean depth has been increasingly used as a demarcation criterion [14]. Distance from shore, water depth, or some combination of the two have been proposed as boundary criteria in the Law of the Sea negotiations.

A relatively broad consensus evolved at the Third United Nations Conference on the Law of the Sea in Caracas in 1974 that coastal states would seek a 12-nautical-mile territorial sea and a 200-nautical-mile outer limit for their exclusive economic zones. More than one hundred countries supported the concept of an economic zone, and there was broad support for coastal states' exclusive economic rights relating to exploration and exploitation of mineral resources. The proposal that coastal-state economic jurisdiction stop at 200 nautical miles is still an active issue, however. Some states with broad coastal margins want to extend their claim to the edge of the margin regardless of distance, while most of the geographically disadvantaged states oppose both the 200-nautical-mile and complete margin proposals. They would prefer to have rights of access to mineral resources of adjacent coastal states. As a compromise, it has been suggested that coastal states be allowed to acquire exclusive rights to the margins but that some portion of the revenue from sale of the output be shared with the international community or with a subset of the international community.

A detailed examination of the issues and alternatives relating to the coastal versus the noncoastal or geographically disadvantaged states will be taken up in Chapter 4. The preceding estimate of offshore production potential and the following estimate of the distribution clearly indicate, however, that the magnitude of the resource pie is very substantial, and therefore that the disposition of offshore hydrocarbon resources is of major concern to the nations of the world. The data in Table 1–4 indicate that if the 200-nautical-mile limit were adopted, the zone would encompass nearly 88 percent of the world's offshore hydrocarbon production potential. As an alternative to a 200-nautical-mile outer limit, adoption of a 3000-meter isobath would slightly increase the proportionate hydrocarbon production potential, whereas a depth criterion of only 200 meters would still result in over two-thirds of total production potential coming under coastal state influence. The National Petroleum Council estimates that about 12 percent of offshore production potential is located within 12 nautical miles from shore [15].

The result is that if coastal states were to acquire exclusive rights to the band of margin between 12 and 200 nautical miles, they would fall heir to the right to acquire approximately 1715 billion barrels of hydrocarbons. This would leave the remaining 12 percent of offshore hydrocarbon production potential to all countries and presumably the first state to claim and exploit a given area would by its action obtain ownership of the hydrocarbon resources.

THE DISTRIBUTION OF OFFSHORE HYDROCARBON POTENTIAL RELATIVE TO NATIONAL INCOME

The distribution of offshore hydrocarbons—petroleum, natural gas and liquid natural gas—among the nations of the world is notably uneven. There are two basic factors accounting for this uneven distribution: a country must have an adjacent coastal margin, and the margin must have a sedimentary formation which contains a hydrocarbon deposit.

Some countries have no continental margins, or they have very small margin areas. This situation may exist because the countries are landlocked, because they possess only short coastlines, because the slope of their continental margin is extremely steep or because they are "zone locked," that is, the offshore zones of bordering states overlap and dominate the offshore area of the first state. Hong Kong and Singapore are primary examples of zone-locked states. While possession of a continental margin area is a prerequisite, the geological nature of the margin as it relates to specific oil-bearing formations is the most crucial variable in determining the distribution of hydrocarbon resources among most countries.

Note carefully that hydrocarbon *potential* is being considered in this cross-country comparison as opposed to hydrocarbon *production potential*, which was discussed in the preceding section. The

Table 1–4. World Subsea Distribution of Discovered and Estimated Ultimate Potential Petroleum[a]
(Total Ocean Area: 362,000,000 sq. km.)

Zone	*Area 10^6 sq. km.*	*Physiographic Province*	*Ultimate Production Potential[b]*	*Share of Total (%)*
Landward of 200 m. Isobath[c]	21.90	Continental shelf	1,544,000	67.98
Seaward of 200 m. Isobath	340.36	Some shelf, almost all slope, all rise and deep ocean basin	728,000	32.04
Landward of 3000 m. Isobath	45.42	All continental shelf and slope, some rise	2,104,800	92.64
Seaward of 3000 m. Isobath	318.15	Most rise, all deep ocean basin	167,200	7.36
Landward of 40 miles from coast	15.66	Mostly continental shelf, some slope	1,350,000	59.42
Seaward of 40 miles from coast	346.87	Some shelf, most slope, all rise and deep ocean basin	922,000	40.58
Landward of 200 miles from coast	77.08	Most shelf and slope, some rise, and deep ocean basins, all small ocean basins	1,988,000	87.50
Seaward of 200 miles from coast	288.04	Some shelf, slope and rise, most deep ocean basin	284,000	12.50

Source: United Nations, General Assembly, Committee on the Peaceful Uses of the Sea-Bed and the Ocean Floor beyond the Limits of National Jurisdiction, *Economic Significance in Terms of Sea-Bed Mineral Resources of the Various Limits Proposed for National Jurisdiction: Report of the Secretary General* (A/AC. 138/87), June 1973.

[a]Expressed in 10^6 barrels of oil. Figures include oil plus oil equivalent of the gas (6000 cubic feet = 1 barrel of oil).

[b]Total estimated ultimately recoverable petroleum, 2272 billion barrels of oil and oil equivalent of the gas. The figures are based on the assumption that all of the petroleum can be economically recovered.

[c]All landward designations include petroleum from shallow sea and small ocean basin shelves and slopes.

United States Geological Survey defines hydrocarbon potential or "potential resources" as "the quantities of crude oil and natural gas that are believed to be present in the sedimentary rocks of unexplored or partly explored regions and that are potentially available through intensive exploration . . . " [16]. An estimate of hydrocarbon potential is a geological statement and does not indicate how much petroleum ultimately will be lifted to the surface. Nevertheless estimates of hydrocarbon potential are very useful in making cross-country comparisons.

Because, in the opinion of the USGS, the use of specific quantity estimates would imply an unwarranted degree of precision, estimates of hydrocarbon potential are presented "in terms of a numerical rating scale that represents a range of more-or-less rough approximations instead of a specific quantity that might imply a misleading degree of precision, accuracy, or certainty" [17]. Each hydrocarbon potential classification encompasses a quantity range which varies by a factor of ten (see Table 1–5 at the end of this chapter). Thus, while there are no numerical estimates of offshore hydrocarbon potential, the USGS ranking provides a sufficient basis to allow extensive international comparisons. A country-by-country listing for offshore oil and natural gas potential is provided in Table 1–5 at the end of this chapter. Table 1–6 is a condensed version of that table.

A review of the data indicates that seventy-nine countries containing a total population of slightly less than three-fourths of the world's population have a mid-level hydrocarbon potential. Within the mid-level group, 53.6 percent of the population has a per capita income of less than $500. The more detailed listing in Table 1–5 indicates that within this group the less-than-$200 per capita income subgroup constitutes the single largest subgroup, with about one-half of the total population. This large concentration of population in a single subcategory is explained by the fact that the People's Republic of China (PRC) and India, as well as several other large population countries (i.e., Indonesia, Nigeria and Bangladesh) have mid-range hydrocarbon potentials. If the PRC and India are excluded from the comparison, the population share of the mid-potential, less-than-$500 per capita category decreases from 53.6 to 27.9 percent of total population. Table 1–5 indicates in brackets other percentage shares allowing for the exclusion of the PRC and India. Only eight countries accounting for 17.8 percent of the total population fall into the high potential category, while the low- and no-potential groups are about evenly matched, with 26 countries (4.5 percent of total population) in the low-potential group and 31 countries (4.1 percent of total population) in the no-potential group.

Table 1–6. National Income Per Capita and Offshore Hydrocarbon Potential for 144 Countries

Hydrocarbon Potential	*Number of Countries*	*Percent of 1972 Total Population*
High (I, II)		
$1200+	6	15.5
$500–$1200	2	2.3
< $500	0	—
Subtotal	8	17.8
Medium (III, IV)		
$1200+	25	12.6
$500–$1200	21	7.4
< $500	33	53.6
Subtotal	79	73.6
Low (V, VI)		
$1200+	5	1.3
$500–$1200	4	0.4
< $500	17	2.8
Subtotal	26	4.5
No Potential		
$1200+	6	1.0
$500–$1200	2	.4
< $500	23	2.7
Subtotal	31	4.1
Total	144	100.0

Note: Table 1–5 indicates that the predominant population share of this group is accounted for by 16 countries with per capita incomes less than $200 and a total population which is 48 percent of the 144–country total.

Using a slightly different approach, the higher income countries account for a total of 30.4 percent of the total 144-country population, with almost one-half of their population share in the high-potential group and a similar proportion in the mid-potential group. The middle-income countries ($500–$1200 per capita) account for only 10.5 percent of total population, most of which is concentrated in the mid-potential category. The next lowest income group ($200–$500 per capita) is the largest population group, with 30.2 percent of total population, and is also concentrated in the mid-potential category. Most of the population in the high-potential category is also in the high income group, whereas most of the population in the no-potential category is in the lowest income group.

It would not be expected that the proportion of population among income groups within each hydrocarbon potential category would correspond to the world distribution of population by income groupings. It is interesting to compare the proportions, however, to indicate the relative degree of "representation" of a given income group within a given potential category. This is done in Table 1–7, with the results shown in the last column. A number near 1 indicates that an income group's proportion of population within a potential category approximates its proportion of the world population. If the number is greater than 1, the representation of the income group is disproportionately greater in that potential category than its proportion in the total population; if it is less than 1, its representation is disproportionately smaller relative to its overall share. It can be seen that the representation of the highest-income subgroup within the high-potential category is disproportionately large as is the representation of the \$500–\$1200 group in the low-potential category and the lowest income subgroup in the no-potential category. Conversely the two lowest income subgroups in the low-potential are well below one while two entries are blank because the particular income group is not represented in the potential category.

A somewhat more methodical examination of the distribution of offshore hydrocarbons has also been undertaken by calculating the correlation coefficients for selected income subgroupings and testing their significance. A total of six comparisons are made and results listed in Table 1–8.

Country per capita income for three country subgroupings are correlated with their respective offshore potentials. In all cases the coefficients tend to be small but positive. A significance ("t") test indicates that the correlation coefficients for the 144-country group and the 107 countries with per capita incomes less than \$1200 are significant at the 99 percent confidence level. The correlation coefficients for the high-income subgroup do not yield similarly high "t-statistics." The highest correlation coefficients are found in the less-than-\$1200 subgroup as are the highest "t-statistics." It may be inferred from these data that there is a loose but statistically significant pattern in which higher income countries tend to have greater offshore hydrocarbon potential and the positive correspondence pattern is strongest in the lower income subgroup. The coefficients for oil and natural gas in each row tend to be similar, reflecting the fact that these two types of hydrocarbons are usually found together (the correlation coefficient between offshore petroleum and offshore natural gas potential is .91).

While the positive signs of the correlation coefficients and their statistical significance are noted, this distributional aspect should not be expected to carry much weight in a policy sense, not only because

Table 1–7. Comparison of Population Share within Each Hydrocarbon Category to Worldwide Population Distribution

Hydrocarbon Potential	*Per Capita Income*	*Percent of Population in Each Potential Category*	*Income Group Total Population as Percent of World Population*	*Index of Representation*
High I, II	$1200+	89.4	30.4	2.94
	$500–$1200	10.6	10.5	1.02
	$200–$500	0	30.2	—
	< $200	0	29.0	—
Middle III, IV	$1200+	20.7	30.4	.68
	$500–$1200	9.0	10.5	.86
	$200–$500	38.8	30.2	1.29
	< $200	31.5	29.0	1.09
Low V, VI	$1200+	24.4	30.4	.80
	$500–$1200	24.9	10.5	2.38
	$200–$500	6.6	30.2	.22
	< $200	44.0	29.0	1.52
None	$1200+	24.6	30.4	.81
	$500–$1200	10.0	10.5	.96
	$200–$500	14.8	30.2	.49
	< $200	50.6	29.0	1.75

Table 1–8. Income per Capita and Offshore Hydrocarbon Potential

	Oil		*Gas*	
	r^a	t	r	t
All Countries (144)	.25	3.09	.22	2.68
Countries with Per Capita Income > $1200 (37)	.11	1.33	.05	.60
Countries with Per Capita Income < $1200 (107)	.35	4.43	.39	5.06

[a] *r* is the coefficient of correlation.

the coefficients are small but, more importantly, because the data are crude and incomplete. It is not inconceivable that the relative positions of the mid- and low-potential countries could improve as the amount of exploration in their offshore areas approaches the amount of exploration which has been conducted off the coasts of many of the industrialized countries.

To focus the examination of distribution more closely on developing economies, the general listing in Table 1–5 is reduced by excluding countries with per capita income greater than $1200, and country classifications are based solely on petroleum potential. Natural gas potential is excluded because its importance to less developed countries (LDCs) as a resource is secondary to that of petroleum [18]. The reduced listing is presented in Table 1–9 and a condensed version of that table is provided in Table 1–10.

As the correlation coefficients would suggest, the distributional pattern for the 101-country subgroup changes only slightly from the 144-country pattern. The middle potential categories, III and IV, still dominate the distribution with 84 percent of the 101-country population. The data in Table 1–10 indicate that the number of countries is distributed relatively evenly in the middle potential category with respect to income, but, as noted in the examination of world distribution, population is heavily concentrated in the lowest income group. Table 1–10 serves to demonstrate the contrast between the two extremes of hydrocarbon potential categories. There is only one country in the entire high-potential range (in category II), and it happens to be relatively wealthy (Mexico). In contrast to this, there are fifty-five countries in the low and no-potential categories, and forty-five of these countries have per capita incomes of less than $500. The entire fifty-five-country subgroup contains only about 13 percent of total Third World population. While the thirty countries in hydrocarbon potential categories V and VI are less fortunate than

Table 1–10. Petroleum Potential Relative to Per Capita Income for 101 Countries *(Summary)*

Petroleum Potential Category		Per Capita Income: $500–$1200 (A) Number of Countries	$500–$1200 (B) Percent of Population	$200–$500 (A) Number of Countries	$200–$500 (B) Percent of Population	< $200 (A) Number of Countries	< $200 (B) Percent of Population	Σ (A)	Σ (B)
High	I	0	—	0	—	0	—	0	—
	II	1	2.02	0	—	0	—	1	2.0
Middle	III	2	1.57	3	6.61	2	25.49	7	33.7
	IV	14	7.94	15	35.91	9	7.73	38	51.6
Low	V	8	2.30	8	.70	13	5.34	29	8.3
	VI	0	—	0	—	1	(< .10)	1	(< .1)
No Potential		2	.58	8	.86	15	2.94	25	4.4
Total		27	14.41	34	44.08	40	4.15	101	100.0

countries with greater hydrocarbon potential, it is still possible that they could realize some direct benefit from offshore exploitation. It is even conceivable that some of the low-potential countries could experience a greater marginal effect than that experienced by countries with larger offshore potentials. There is, however, no certain means to determine this outcome in advance.

There is, however, one clear-cut case in which no direct benefit can be expected. There are twenty-five LDCs which have zero offshore petroleum potential and only two of them have a per capita income exceeding $500. Relative to the other country subgroups, the no-potential countries are the poorest while comprising a relatively small share of Third World population(about 4.4 percent, or approximately 110 million inhabitants). Of the fifteen countries with per capita income less than $200 in the no-potential category, all are classified by the United Nations as least developed countries [19].

Overall, the distributional pattern of offshore petroleum potential may be characterized as one in which most of the nations in the world (with most of the world's population) have some potential for offshore hydrocarbon exploitation. About one-half of the group of 101 countries with per capita income less than $1200 have a moderate potential, one-fourth have a small potential and one-fourth, including most of the poorest nations in the world, have no potential.

It has been shown that for 144 countries there is a positive and significant correspondence between national income per capita and offshore hydrocarbon potential, but this finding was accompanied by a caveat on the meaning and dependability of the coefficients. Even allowing for inaccuracies, however, it is significant that the distributional pattern alone does not provide a basis for the usual Industrial World-Third World debate. But the fact that a small minority of the world's population, which is also at a subsistence income level, has no promise of ever realizing direct economic advantage from offshore hydrocarbon resources does give rise to an issue. The real division is between this group and the group of nations which have a reasonable expectation of offshore hydrocarbon production. From the standpoint of distribution, if there is to be a debate over division of the benefits from offshore hydrocarbon exploitation, the discussion should focus on the actual distributional pattern and recognize the fact that the petroleum "have-nots" are the same small group which have benefited least from the distribution of the world's other natural resources and from postwar economic growth.

NOTES TO CHAPTER 1

1. The areal extent of the continental margin is not exactly known. The area used to represent the continental margin is the area encompassed by the shore and a line 200 nautical miles from the shore, which for 121 countries is 25.3 million square nautical miles. The margin area extending to a uniform water depth of 3,000 meters is 13.7 million square nautical miles. (U.S. Department of Interior, United States Geological Survey, *Summary of 1972 Oil and Gas Statistics for Onshore and Offshore Areas of 151 Countries*, by Sherwood E. Frezon, Geological Survey Professional Paper 885 [Washington, D.C., 1974], p. 156).

2. National Petroleum Council, *Petroleum Resources under the Ocean Floor* (Washington, D.C.: National Petroleum Council, March 1969), p. 24.

3. Don E. Kash *et al.*, *Energy under the Oceans: A Technology Assessment of Outer Continental Shelf Oil and Gas Operations* (Norman: University of Oklahoma Press, 1963), pp. 11, 14 and 310.

4. Ibid., pp. 310–11.

5. U.S. Department of Interior, USGS, *Summary of 1972 Oil and Gas Statistics*, p. 2.

6. About 3.3 billion barrels of oil were produced from the offshore area in 1972 using the broad interpretation of offshore. This is equivalent to 18 percent of total world production.

7. U.S. Department of Interior, USGS, *Summary of 1972 Oil and Gas Statistics*, p. 162.

8. National Petroleum Council, *Petroleum Resources*, p. 24.

9. U.S. Atomic Energy Commission, Oak Ridge National Laboratory, *Survey of Energy Resources: 1974*, ed. William G. Peck (New York: U.S. National Committee of the World Energy Conference, 1974), p. 93.

10. United Nations, General Assembly, Committee on the Peaceful Uses of the Sea-Bed and the Ocean Floor beyond the Limits of National Jurisdiction, *Economic Significance in Terms of Sea-Bed Mineral Resources of the Various Limits Proposed for National Jurisdiction: Report of the Secretary-General* (A/AC. 138/87), 4 June 1973, p. 11.

11. The estimates by Lewis G. Weeks are the most widely cited and the most recent. Lewis Weeks is a former chief geologist for Exxon and president of the American Association of Petroleum Geologists, who has served as an adviser to various governments and to the United Nations. He was president of Oil Basins Ltd., Offshore Royalties, Inc., and Weeks Natural Resources, and has co-authored several books on petroleum as well as having written numerous technical papers and articles. *Who's Who in America*, p. 3240.

12. Kash *et al.*, *Energy under the Oceans*, p. 314.

13. The ocean depth criterion is usually expressed as an "isobath," which can be conceived of as a vertical plane between the seabed and the ocean surface which follows the contours of the ocean floor so that its height remains constant as a path is traced along the ocean floor.

14. UN Committee on Peaceful Uses, *Economic Significance of Limits*, p. 15.

15. National Petroleum Council, Committee on Ocean Petroleum Resources, *Ocean Petroleum Resources: An Interim Report of the National Petroleum Council* (Washington, D.C.: National Petroleum Council, July 4, 1974), p. 4.

16. U.S. Department of Interior, United States Geological Survey, *Summary Petroleum and Selected Mineral Statistics for 120 Countries, Including Offshore Areas*, by John P. Albers *et al.*, Geological Survey Professional Paper 817 (Washington, D.C., 1973), p. 2.

17. *Ibid.*, p. 3.

18. Although petroleum and natural gas categories are based on similar numerical values (e.g., 10,000 billion barrels of oil or 10,000 trillion cubic feet of gas), these amounts are not equivalent in terms of energy, gross revenue potential or net revenue potential. In addition natural gas requires extraordinarily large technical and investment input. As a result, natural gas is not likely to be produced and sold in major quantities from LDCs in their early stages of offshore development.

In energy terms, 6000 cubic feet of gas is equivalent to one barrel of oil. Thus, 100 trillion cubic feet of natural gas, the lower limit in category II, is equivalent to 16.7 billion barrels of oil, in contrast to the lower limit for oil in category II of 100 billion barrels. In revenue terms the price of oil is approximately $10 per barrel, whereas the price of 1000 cubic feet of natural gas is $1. Again, citing the lower limits of category II for oil and gas, the gross value of 100 billion barrels of oil is $1 trillion but the gross value of 100 trillion cubic feet of gas is only $100 billion.

Besides these differences, the transportation and distribution of natural gas requires a technically sophisticated system because of its volatile nature, and there is a very large minimum initial investment requirement because of the physical nature and price of natural gas. If the gas is to be transported overland, it is possible to move it in its gaseous state in large-diameter pipe (assuming there is an end user, such as an industrial complex or a distribution system for individual users). If the gas is to be transported by sea, it is necessary to liquify the gas and carry it in insulated containers. Again, the minimum viable scale of the operation is very large.

Plans to ship natural gas from Nigeria to the United States illustrate the magnitude of a transoceanic shipping operation. In Nigeria approximately 1000 to 2000 cubic feet of natural gas is flared for each barrel of oil produced. In 1974 Nigerian daily oil output averaged 2.3 million barrels; thus approximately 839.5 billion cubic feet of natural gas was wasted. Nigeria intends to invest $2.2 billion in two natural gas liquification plants and to use the services of ten to twelve ships with insulated containers (cyrogenic tankers) to transport the gas to the United States. A single cyrogenic tanker costs about $50 million. The undertaking, estimated to cost about $3.3 billion, will accommodate only a portion of total Nigerian natural gas output. Henry S. Hayward, "Nigeria's Costly 'Flaring' of Gas." *Christian Science Monitor*, 24 October 1974.

In the actual production process, natural gas is frequently brought to the surface along with petroleum, but only because petroleum cannot readily be recovered alone. Routinely the two hydrocarbons are separated at the wellhead; then the natural gas is burned or "flared." In some instances it may be reinjected. In production areas that are remote from markets, natural gas tends to be an unused byproduct because there are no transportation and storage facilities. Thus much of the natural gas produced in the Middle East is flared, but the gas deposits on the North Sea, which are close to the large European markets, and those in the American Gulf Coast are being exploited. As there occur technical advances, price increases and development of nearby markets for natural gas, this commodity can be expected to become an important economic resource in LDC offshore production.

19. The Committee for Development Planning of the United Nations uses three criteria to determine the least developed economies, sometimes referred to as relatively less developed economies (RLDCs). The criteria are: (a) annual per capita gross domestic product is $100 or less; (b) the manufacturing sector represents 10 percent or less of GDP; and (c) the literate population represents 20 percent or less of the total adult population (age above 15). These criteria constitute essentially a minimum level for classification as an RLDC; thus, some economies which do not fit all three criteria but nevertheless exhibit the classic characteristics of an economy at the extreme low end of the development scale are included. The Central African Republic is an example, it is considered to be an RLDC although its share of manufacturing exceeds 10 percent of a total gross domestic product. The total population of the least developed countries is approximately 200 million. As of August 1975, 29 less developed nations were classified as RLDCs by the UN.

Table 1–5. Distribution of Offshore Hydrocarbon Potential Relative to Per Capita Income for 144 Countries

1974 GNP per Capita	*Country*	*Potential Ultimate Recoverable Reserve Category*[1] *Oil*	*Gas*	*Population*[2]	*Population as Percent of Total*
Countries with High Potential for Recoverable Reserves (I, II)					
$2500+				336,975	8.63
	Australia	III	II	124,415	(13.41)**
	Libya	II	II	2,160	
	United States	II	I	210,400	
$1200–$2500				268,774	6.88
	Saudi Arabia	II	II	7,745	(10.70)
	USSR	II	II	249,750	
	Venezuela	III	II	11,279	
$500–$1200				88,183	2.26
	Iran	III	II	32,136	(3.51)
	Mexico	II	III	56,047	
< $500	—			—	—
Countries with Medium Potential for Recoverable Reserves (III, IV)					
$2500+				357,310	9.15
	Belgium*	IV	IV	9,760	(14.22)
	Canada	III	III	22,130	

Table 1–5. cont'd overleaf . . .

Table 1–5. continued

1974 GNP per Capita	*Country*	*Potential Ultimate Recoverable Reserve Category*[1]		*Population*[2]	*Population as Percent of Total*
		Oil	*Gas*		
	Democratic Republic of Zaire	V	IV	16,980	
	Denmark	IV	IV	5,020	
	Federal Republic of Germany	IV	IV	61,970	
	France	IV	IV	52,160	
	Israel*	IV	IV	3,210	
	Japan	IV	IV	108,350	
	Kuwait*	III	IV	880	
	Netherlands	IV	III	13,430	
	New Zealand	V	III	2,960	
	Norway	III	III	3,960	
	Qatar	III	III	180	
	United Arab Emirates	III	III	320	
	United Kingdom	III	III	56,000	
$1200–$2500				136,853	3.50
	Argentina	III	III	24,282	(5.45)
	Brunei*	IV	V	145	
	Bulgaria	IV	IV	8,622	
	Cyprus	IV	II	635	
	Gabon	IV	IV	520	
	Greece	IV	IV	8,930	
	Ireland	V	IV	3,030	
	Italy	IV	III	54,890	
	Spain	IV	V	34,740	
	Trinidad & Tobago	IV	IV	1,059	

$500–$1200				290,349	7.44
	Bahrain*	IV	IV	232	
	Brazil	IV	III	101,051	
	Chile	IV	III	10,230	
	Costa Rica	V	IV	1,872	
	Cuba	V	IV		
	Iraq*	IV	IV	10,410	
	Jamaica	IV	IV	1,967	
	Lebanon	IV	IV	2,977	
	Malaysia	III	III	11,302	
	Malta	IV	IV	320	
	Nicaragua	IV	IV	1,973	
	Oman	IV	IV	720	
	Panama	IV	IV	1,570	
	Peru	IV	IV	14,531	
	Republic of China	IV	IV	15,424	
	Romania	IV	IV	20,700	
	Turkey	IV	IV	39,930	
	Union of South Africa	V	IV	24,255	
	Uruguay	V	IV	2,995	
	Yugoslavia	V	IV	20,960	
$200–$500				1,146,170	29.35
	Albania	IV	IV	2,295	
	Angola	IV	IV	5,717	(13.33)
	Cameroons*	V	IV	6,206	
	Colombia	IV	III	22,500	
	Democratic Republic of Korea	IV	IV	15,040	
	Ecuador	IV	III	6,786	
	Egypt	III	III	35,619	
	Ghana	IV	IV	9,313	
	Guatemala*	V	IV	5,175	

Table 1–5. *cont'd overleaf . . .*

Table 1–5. continued

1974 GNP per Capita	Country	Potential Ultimate Recoverable Reserve Category[1]		Population[2]	Population as Percent of Total
		Oil	Gas		
	Honduras	IV	IV	2,781	
	Ivory Coast	IV	IV	5,887	
	Morocco	IV	IV	15,903	
	Mozambique	IV	III	8,283	
	Nigeria	III	III	71,262	
	People's Republic of China	IV	III	811,350	
	Philippines	IV	III	40,219	
	Republic of Korea	III	III	32,905	
	Senegal	IV	IV	4,070	
	Thailand	IV	IV	39,400	
	Tunisia	IV	IV	5,459	
< $200				947,173	24.25
	Arab Republic of Yemen*	IV	IV	6,217	(14.54)
	Bangladesh	IV	III	94,000	
	Burma	IV	III	29,509	
	Ethiopia	V	IV	7,343	
	Guinea	V	IV	5,243	
	India	III	III	581,911	
	Indonesia	III	III	124,415	
	Khmer Republic	IV	IV	7,343	
	Malagasy Republic	IV	IV	8,301	
	Mauritania	IV	IV	1,257	
	Pakistan	IV	IV	66,230	
	People's Democratic Republic of Yemen	IV	IV	1,560	
	Republic of Vietnam	IV	IV	19,870	

	Tanzania	V	IV	13,974	
Countries with Low Potential for Recoverable Reserves (V, VI)					
$2500+				12,800	.33
	Finland	V	V	4,660	(.51)
	Sweden	V	V	8,140	
$1200–$2500				35,735	.92
	Bahamas	VI	VI	190	(1.42)
	Poland	V	V	33,360	
	Singapore*	V	V	2,185	
$500–$1200				13,803	.35
	Barbados*	V	V	239	(.55)
	Belize*	V	V	132	
	Dominican Republic	V	V	4,438	
	Portugal	V	V	8,994	
$200–$500				8,017	.21
	El Salvador	V	V	3,771	
	Equatorial Guinea	V	V	313	
	Guinea-Bissau	V	V	510	
	Guyana	V	V	772	
	Liberia	V	V	1,452	
	People's Republic of the Congo*	V	V	1,199	
< $200				102,284	2.62
	Benin* (Dahomey)	V	V	2,947	(4.07)
	Democratic Republic of Vietnam	V	V	23,220	

Table 1–5. *cont'd overleaf . . .*

Table 1–5. continued

1974 GNP per Capita	*Country*	*Potential Ultimate Recoverable Reserve Category*[1]		*Population*[2]	*Population as Percent of Total*
		Oil	*Gas*		
	Gambia, The	V	V	493	
	Haiti	V	V	4,454	
	Kenya	V	V	12,480	
	Maldives*	VI	VI	114	
	Sierra Leone	V	V	2,787	
	Sri Lanka	V	V	13,180	
	Sudan	V	V	17,051	
	Togo	V	V	2,120	
	Zaire	V	V	23,438	
Countries with No Potential for Recoverable Reserves					
$2500+				6,990	.18
	Iceland	—	—	210	(.28)
	Luxembourg	—	—	350	
	Switzerland	—	—	6,430	
$1200–$2500				32,532	.83
	Austria	—	—	7,530	(1.30)
	Czechoslovakia	—	—	14,572	
	Hungary	—	—	10,430	

$500–$1200				16,057	
	Algeria			14,700	.41
	Mongolia			1,357	(.64)
$200–$500				23,845	.61
	Bolivia	—	—	5,331	(.95)
	Botswana	—	—	641	
	Jordan	—	—	2,540	
	Mauritius	—	—	860	
	Paraguay	—	—	2,416	
	Swaziland	—	—	463	
	Syria	—	—	6,948	
	Zambia	—	—	4,646	
< $200				81,341	2.08
	Afghanistan	—	—	16,634	(3.24)
	Bhutan	—	—	1,124	
	Burundi	—	—	3,580	
	Central African Republic	—	—	1,710	
	Chad	—	—	3,873	
	Laos	—	—	3,110	
	Lesotho	—	—	1,165	
	Malawi	—	—	4,833	
	Mali	—	—	5,370	
	Nepal	—	—	12,020	
	Niger	—	—	4,357	
	Rwanda	—	—	3,980	
	Somalia	—	—	3,042	
	Uganda	—	—	10,829	
	Upper Volta	—	—	5,714	

Table 1–5. cont'd overleaf . . .

Table 1–5. Source and Notes

Source: Geological Survey, *Summary of 1972 Oil and Gas Statistics for Onshore and Offshore Areas of 151 Countries*: IBRD, *World Bank Atlas*, 1975.

*Margins less than 10,000 square nautical miles at 3000-meter depths.

**Numbers in parentheses are percentages excluding the People's Republic of China and India.

1, Potential Resources are quantities of crude oil and natural gas that are believed to be present in partially explored areas. The quantities of oil and gas assigned to the various categories are the following:

Category	*Quantity of Oil (Billions of barrels)*	*Quantity of Gas (Trillions of Cubic Feet)*
I	1000–10,000	1000–10,000
II	100–1000	100–1000
III	10–100	10–100
IV	1–10	1–10
V	0.1–1	0.1–1
VI	0.01–0.1	0.01–0.1
0	0.0–< 0.01	0.0–< 0.01

2. World total is the sum of the countries included in the table (3,905,191,000); excluding India and the People's Republic of China, the total is 2,511,930,000.

Table 1–9. Petroleum Potential Relative to Per Capita Income for 101 Countries

Petroleum Potential Category	*Per Capita Income*	*Country*	*Population ('000)*	*Population as Percent of Total World Population*
I	—	—	—	—
II	$500–$1200	Mexico	56,047	
			56,047	2.02
II	< $500	—		—
III	$500–$1200	Iran	32,136	
		Malaysia	11,302	
			43,438	1.57
III	$200–$500	Egypt	35,619	
		Korea	32,905	
		Nigeria	71,262	
			183,224	6.61
III	< $200	India	581,911	
		Indonesia	124,415	
			706,326	25.49
IV	$500–$1200	Bahrain	232	
		Brazil	101,051	
		Chile	10,230	
		China, Republic of	15,424	
		Iraq	10,410	
		Jamaica	1,967	
		Lebanon	2,977	
		Malta	320	

Table 1–9. cont'd overleaf . . .

Table 1–9. continued

Petroleum Potential Category	*Per Capita Income*	*Country*	*Population ('000)*	*Population as Percent of Total World Population*
		Nicaragua	1,973	
		Oman	720	
		Panama	1,570	
		Peru	14,531	
		Romania	20,700	
		Turkey	37,930	
			220,035	7.94
IV	$200–$500	Albania	2,295	
		Angola	5,717	
		Colombia	22,500	
		Democratic Republic of Korea	15,040	
		Ecuador	6,786	
		Ghana	9,313	
		Honduras	2,781	
		Ivory Coast	5,887	
		Morocco	15,903	
		Mozambique	8,283	
		People's Republic of China	811,350	
		Philippines	40,219	
		Senegal	4,070	
		Thailand	39,400	
		Tunisia	5,459	
			995,003	35.91
IV	< $200	Arab Republic of Yemen	6,217	
		Bangladesh	74,000	
		Burma	29,509	

		Khmer Republic	7,343	
		Malagasy Republic	8,301	
		Mauritania	1,257	
		Pakistan	66,230	
		People's Democratic Republic of Yemen	1,560	
		Republic of Vietnam	19,870	
			214,287	7.73
V	\$500–\$1200	Barbados	239	
		Belize	132	
		Costa Rica	1,872	
		Cuba	8,920	
		Dominican Republic	4,438	
		Union of South Africa	24,265	
		Uruguay	2,995	
		Yugoslavia	20,960	
			63,820	2.30
V	\$200–\$500	Cameroons	6,206	
		Dominican Republic	4,438	
		El Salvador	3,771	
		Equatorial Guinea	313	
		Guatemala	5,175	
		Guinea-Bissau	510	
		Guyana	772	
		Liberia	1,452	
		People's Republic of the Congo	1,199	
			19,398	.70
V	< \$200	Benin (Dahomey)	2,947	
		Democratic Republic of Vietnam	23,220	

Table 1–9. cont'd overleaf . . .

Table 1–9. continued

Petroleum Potential Category	*Per Capita Income*	*Country*	*Population ('000)*	*Population as Percent of Total World Population*
		Ethiopia	26,550	
		Gambia, The	493	
		Guinea	5,243	
		Haiti	4,454	
		Kenya	12,480	
		Sierra Leone	2,787	
		Sri Lanka	13,180	
		Sudan	17,051	
		Tanzania	13,974	
		Togo	2,120	
		Zaire	23,438	
			147,937	5.34
VI	> $200	—		—
VI	< $200	Maldives	114	
			114	0.0
No Potential	$500–$1200	Algeria	14,700	
		Mongolia	1,357	
			16,057	.58
No Potential	$200–$500	Bolivia	5,331	
		Botswana	641	
		Jordan	2,540	
		Mauritius	860	
		Paraguay	2,416	
		Swaziland	463	

		Syria	6,948	
		Zambia	4,646	
			23,845	.86
No Potential	< $200	Afghanistan	16,634	
		Bhutan	1,124	
		Burundi	3,580	
		Central African Republic	1,710	
		Chad	3,873	
		Laos	3,110	
		Lesotho	1,165	
		Malawi	4,833	
		Mali	5,370	
		Nepal	12,020	
		Niger	4,357	
		Rwanda	3,980	
		Somalia	3,042	
		Uganda	10,829	
		Upper Volta	5,714	
			81,341	2.94

Source: International Bank for Reconstruction and Development, *World Bank Atlas: 1975* (Washington: IBRD, 1975); Geological Survey, *Summary of 1972 Oil and Gas Statistics for Onshore and Offshore Areas of 151 Countries.*

Chapter 2

Aspects of Offshore Hydrocarbon Exploitation

INTRODUCTION

This chapter focuses on three aspects of offshore hydrocarbon exploitation: technical capability, costs and commercial viability. The technical capability to exploit offshore hydrocarbons in coastal waters less than 100 meters deep is an operational reality. Exploitation beyond 100 meters is relatively rare, and production has not yet been extended beyond 200 meters. However, the technology to extract hydrocarbons from beneath more than 200 meters of water is now becoming available.

While there appears to be a ready technical capacity to produce oil from great water depths and under a large variety of conditions, discovery and production of seabed hydrocarbons require a very large initial investment. The investment requirement increases rapidly as operations are extended beyond 200 meters. Thus the issue of commercial viability becomes crucial for offshore operations beyond 200 meters.

The technical feasibility as well as the commercial viability of offshore operations are, of course, of interest to all developing nations. But it should be noted that the poor, non-coastal countries have a particular interest in exploitation of the more seaward portions of the continental margins, since it is from this portion of the continental margins that these states are likely to have the greatest opportunity to obtain some direct benefit.

DEVELOPMENT AND EXPLOITATION OF OFFSHORE HYDROCARBONS

The ultimate total amount of subseabed hydrocarbons and their location are dependent on geological and physical phenomena. The total amount and the distribution are fixed and immutable. An economic examination, however, focuses on hydrocarbons as economic resources, that is, on the amount and the distribution of those hydrocarbon resources, the locations of which are known and which are accessible for extraction. Clearly, knowledge of location is necessary, but the problem of accessibility must also be solved before the hydrocarbons can shift from a geological fact to an economic factor. This differentiation between offshore hydrocarbon formations as geological fact and offshore hydrocarbons as economic resources is significant because the amount and the distribution of hydrocarbons as economic resources can change. Changes can occur as frequently as continuing exploratory efforts and technological improvements lead to new discoveries and as technical advances increase capability to extract the hydrocarbons.

Historically, constant exploration efforts and technical improvements have altered the distribution of known offshore hydrocarbons. As a result of technical advances, the distribution of offshore hydrocarbons has spread more evenly throughout the world. It is, therefore, useful to have an idea of how subsea hydrocarbons are found and produced and to be aware of the significance of the role of technical development as it affects offshore hydrocarbon distribution.

EXPLORATION, DEVELOPMENT AND PRODUCTION

Exploration, the first step in exploitation of offshore hydrocarbon resources, is usually carried out in two phases. In the first phase, sometimes called passive reconnaissance, large areas are mapped by air or seaborne instruments which take magnetic readings and which measure local variations in the earth's gravity. The second phase may include such methods as seismic surveying and recovery of soil samples from the ocean floor [1]. Exploration capability has generally been able to exceed production ability. Water depth, wave amplitude, winds, currents and nature of the sea floor are not critically important elements in passive exploration activity, but these factors critically influence the second step in offshore hydrocarbon exploration—exploratory drilling.

At the outset of offshore operations, it was apparent that the cost of drilling an offshore well in an unproved area (commonly referred to as a "wildcat" well) from a fixed platform was excessive relative to the probable return. Thus, the development of offshore mobile drilling units or "rigs" became a key link in the effort to seek offshore hydrocarbon deposits [2]. As an example of present exploratory capability, in 1973 a well was drilled in 396 meters (1300 feet) of water off northwest Borneo using a dynamically positioned drill ship (i.e., no mooring lines) [3].

While the oil industry was pursuing development and expansion of underwater exploratory drilling capacity, scientifically motivated reconnaissance work was simultaneously expanding knowledge of the earth's crust under the ocean and developing technical advances in underwater drilling techniques. A major effort in this area was conducted under the auspices of the Joint Oceanographic Institutions for Deep Earth Sampling (JOIDES). By 1974 the JOIDES project had succeeded in sampling the crust of oceanic sediments at water depths of 6207 meters (20,365 feet) and—a major accomplishment—the JOIDES drill ship had reentered holes as deep as 5757 meters (18,888 feet) below the ocean's surface [4].

To summarize, it is now feasible, using the most advanced offshore drilling technology, to conduct exploratory drilling at a 1000-meter (3281-foot) depth. Equipment ordered in 1974 had a design capacity to allow drilling and completion at water depths of 1800 meters (5906 feet). Drill ships can be moored in as much as 1000 meters (3281 feet) of water, and there is virtually no water depth limit if dynamic positioning is used. Undersea blowout preventers and control systems necessary for drilling have a depth capability of 3000 meters (9900 feet).

The last major step in the production of offshore hydrocarbons is the drilling of development wells and installation of "completion" or production equipment. After the initial wildcat or exploratory well and subsequent tests are completed, including the drilling of additional wells to determine the size and nature of the find and the optimum location for a production platform, a platform is set in place and a number of development wells are drilled. The appropriate production equipment is installed to control output, thereby establishing the final production of crude oil and/or natural gas. The basic component in the development and production phase is the fixed platform, which consists of steel platforms mounted on steel pilings driven into the ocean floor or situated on steel legs which rest on the ocean floor.

Fixed platforms are used in the production phase instead of mobile units because production is likely to continue for several years and fixed platforms have much lower daily operating costs than mobile units. Most drilling production platforms now have a 12- to 25-well capacity, and one platform in the North Sea, reportedly the most expensive to date at a cost of over $100 million, will have 27 wells. In addition, this three-deck structure will contain all handling equipment for gas and oil as well as transfer pumps for a platform-to-shore pipeline [5]. A 32-well platform design has been proposed for use in the Mediterranean [6].

Since the first platform out of sight of land was constructed in 1947 in the Gulf of Mexico at a water depth of approximately 50 feet, constant development has increased the operational scope of platforms to enable them to function under such diverse conditions as those found in the North Sea, the Cook Inlet of Alaska and the waters of the Indonesian archipelago. Advances in the structural design of platforms now permit operation in water depths up to 167 meters (550 feet), and in one case a platform has been constructed in 257 meters (850 feet) of relatively calm water in the Santa Barbara channel off the southern California coast [7].

One of the most recent advances in the fixed platform concept has been designed by an Italian-based joint enterprise, Tecnomore S.P.A. The innovative gravity design of this platform promises to effect substantial cost reductions relative to conventional platforms. The platform has a 32-well capacity and two drilling rigs, and can operate in 198 meters (650 feet) of water. This design represents an excellent example of the constant effort to reduce platform structure and installation costs and to increase platform productivity by combining as many operations as possible on one platform while maximizing the number of wells drilled [8].

While there are significant technical problems in building platforms for greater depths, it has been suggested that a 305-meter (1000-foot) depth is technically possible. Even before the 305-meter technical depth limit is reached, however, the cost of this type of platform is thought to be prohibitive [9]. A major reason behind the increase in the number of wells per platform and the increasing number of functions the platform can accommodate, has been to increase the overall productivity of the individual platform. This increase in productivity has in turn been instrumental in permitting the more costly deep water platform implacements.

The subsea completion system is one of the recent and most promising responses to the high cost of producing hydrocarbons in deep, rough waters. Using this system, development wells are drilled by mobile units; then all the necessary wellhead equipment, including the blowout preventer, is connected to the well on the ocean floor. Output from the well is transferred via manifolds through pipelines to a platform, surface ship, shore facility or offshore storage facility. Subsea completion systems may be either remotely controlled wet units or they may consist of conventional surface equipment placed in a capsule filled with some inert gas such as nitrogen [10].

Subsea systems tend to be less sensitive to climate-oceanographic conditions and water depth. In addition, such systems tend to be suited for smaller reserve volumes and larger areas of accumulation. Because they can be more widely dispersed over the accumulation area, relatively fewer subsea wells are required to develop a given volume of reserves [11].

The subsea completion concept is probably the most advanced approach for offshore petroleum production, but it is not an entirely new or untested concept. In 1966 more than 100 subsea completion systems were in operation [12]. In United States offshore waters, 70 subsea completions were in use in 1973 [13].

A major point to be drawn from the preceding discussion is that the fundamental problem of offshore hydrocarbon inaccessibility is essentially solved. A substantial portion of the world's offshore hydrocarbon resources are technically accessible for exploitation with the possible exception of some polar areas. This is not to say that all offshore hydrocarbon deposits are commercially viable or that there is no backlog of items waiting for technical development and refinement. The remaining technical development backlog, however, consists primarily of advances which will allow reductions in cost, increases in reliability—in some cases virtually to a fail-safe condition—and decreases in equipment redundancy. The question of commercial viability for any specific offshore hydrocarbon deposit remains dependent on such basic variables as the price of hydrocarbons, the amount of ultimately recoverable resources from the given deposit and the precise state of the technical art at the time.

THE FINANCIAL COST AND COMMERCIAL VIABILITY OF OFFSHORE OPERATIONS

The presence of hydrocarbons offshore and the technical ability to extract them from beneath the ocean floor are not sufficient conditions for their production. It is also necessary that the estimate of the present value of the income stream including an allowance for risk be greater than the estimated present value of expenditures.

The costs associated with offshore production tend to be substantially higher than onshore production costs. Obviously, costs can vary substantially with local conditions; however, in comparing an offshore oil field in 20 meters of water with a similar onshore oil field, the offshore oil field investment is approximately three and a half times greater than that required for the onshore field. If all of the ancillary facilities are included, the offshore investment is about twice the onshore investment. An offshore gas field in 70 meters of water in the North Sea and its ancillary facilities require an investment about four times greater than that for a comparable onshore operation [14].

The cost of offshore exploitation also increases with increases in water depth. It was estimated in 1967 that for a model field in the Gulf of Mexico the capital investment doubled as water depth increased from 30 meters to 180 meters, and the investment more than quadrupled for a water depth of 300 meters [15]. The total investment requirement for a fourteen-well offshore oil complex in 20 meters of water in the Gulf of Mexico was estimated in 1968 to be about $39 million. An offshore three-platform natural gas complex in 70 meters of water in the North Sea required a total investment of $69.4 million in 1968 [16]. According to a recent estimate, the Ekofisk field complex in the North Sea will involve total capital costs of $4 billion [17]. To date, most offshore hydrocarbon development has taken place at water depths less than 70 meters. There are exceptions: oil discoveries have been made at water depths greater than 400 meters, and one production platform has been constructed in the Santa Barbara channel at a water depth just exceeding 200 meters.

While offshore operations are markedly more expensive than onshore, it is reasonably certain that costs at water depths less than 200 meters will not be sufficiently high to inhibit seriously wide-scale exploitation in this shallower seabed area. It is interesting in the general case, however, to consider what the outer limit of exploitation with respect to water depth and other factors might be. From the viewpoint of geographically disadvantaged LDCs, it is particularly relevant to examine the financial viability of exploitation in the 340-million-square-kilometer area of margin seaward of the 200-meter isobath [18].

The 200-meter depth is an appropriate starting point because it was recognized as a principal seabed demarcation line between national and international jurisdications even prior to the first Law of the Sea Convention in 1958. Furthermore, the seabed area beyond 200 meters water depth has been specified in many documents, including the 1970 United States working paper, Draft United Nations Convention on the International Seabed Area, as the area from which geographically disadvantaged states might realize some benefits. This area seaward of the 200-meter isobath—referred to as "the deep offshore, comprising the outer continental shelf, the continental slope, and the adjacent oceanic regions beyond"—is thought to have an ultimate production potential of over 700 billion barrels of oil, or about 23 percent of the world total offshore production potential [19]. The basic questions which evolve are whether hydrocarbon exploitation costs can be expected to be consistently less than expected returns, so that hydrocarbon production from the deep offshore area will be viable and, if production is viable for some of the deep offshore area, what are the limits?

Table 2–1 provides some cost estimates for offshore exploitation relative to water depths between 200 and 1000 meters and varying climate-oceanographic conditions. The estimates in Table 2–1 represent typical costs; therefore, each figure should be interpreted as a mid-point within a range of possible costs.

The cost estimates are presented in two columns, cost per well and cost for five wells. The latter figure is provided so total costs of a production complex will more closely reflect the sum of exploration costs necessary to establish production. The quintupling of exploration costs implies a success ratio of one discovery out of five "new-field wildcat" wells [20] for a 20 percent success ratio [21].

The cost estimates for exploration are understated to the extent that they do not include related exploratory expenditures such as lease bonuses and other overhead costs associated with acquisition of exploration rights. The data indicate that exploratory costs approximately double as operations move from the 500-meter depth to the 800-meter depth. This increase occurs because it becomes necessary to use the more expensive semisubmersible drilling platforms or dynamically positioned drill ships. If an exploratory well were to be drilled, for example, in the Straits of Magellan, the cost might be expected to be close to $25 million (200 meters in severe conditions).

Development well drilling and installation of production equipment constitute the next major step after the conclusion of a successful exploratory operation. Table 2–2 provides cost estimates of typical development and production expenditure requirements (exploration costs are not included). The cost estimates in Table 2–2 are based on a system which would include two platforms of 20 wells each and production facilities sufficient for a 25,000-barrel-per-day output. It is further assumed that subsea well depth is 2424 meters (8000 feet). Beyond the 200-meter water depth, the probability of

Table 2–1. Cost of Offshore Exploration *(In 1974 Millions of Dollars)*

Water Depth		*Climate*					
		Mild[a]		*Moderate*[a]		*Severe*[a]	
Meters	*Feet*	*Cost/Well*	× 5	*Cost/Well*	× 5	*Cost/Well*	× 5
200	660	2.16	10.80	2.70	13.50	4.86	24.30
500	1650	2.70	13.50	3.51	17.55	5.67	28.35
800	2640	6.21	31.05	7.02	35.10	8.91	44.55
1000	3300	6.41	32.05	7.22	36.10	9.11	45.55

Source: Adapted from Table 3, *Ocean Petroleum Resources: An Interim Report of the National Petroleum Council*, July 1974.

[a]Typical of the various climatic conditions are:
Mild: Senegal, Gabon, Honduras, Mediterranean, Java Sea, Persian Gulf;
Moderate: Gulf of Mexico, South Atlantic, South Pacific,* Northwest Australia,* Sea of Japan,* Yellow Sea;
Severe: North Sea, Bay of Biscay, South Australia, Gulf of Alaska,* North Atlantic, North Pacific, West Coast of Canada, Nova Scotia.+

*Earthquake area. +Iceberg area.

Table 2–2. Development and Production Expenditure Requirements
(In 1974 Millions of Dollars)

Water Depth		*Climate*[a]		
Meters	*Feet*	*Mild*	*Moderate*	*Severe*
200	660	85.5	95.0	266.0
500	1650	256.5	285.0	n/a[b]
1000	3300	408.5	456.0	969.0

Source: Adapted from Table 4, *Ocean Petroleum Resources: An Interim Report of the National Petroleum Council*, July 1974.
[a]See Table 2–1 for examples of climate localities. [b]n/a = not available.

employing a subsea completion system could be expected to increase, with an even greater increase occurring as climate-oceanographic conditions become more severe.

While the cost estimates in Table 2–2 pertain to platforms, it is not certain that platforms will be widely used in practice. The conclusions from one comparative analysis of subsea systems and platforms suggests that a subsea completion system may yield a higher present value profit. (The subsea completion system used in the comparison is a dry system developed by Lockheed Petroleum Services, Ltd.) The investment requirement for a platform in 100 feet of water was estimated to be 1.8 times greater than a comparable subsea completion system, 2.6 times greater in 200 feet of water, and 4 times greater in 300 feet of water. Operating costs were similar but the platform system required a minimum of thirty-nine months from start of the first delineation well to commencement of production, whereas about fourteen months were necessary for a subsea completion system. Because of these factors and the significant fact that subsea units are more divisible than platforms in terms of investment, subsea systems tend to be more advantageous as higher discount rates are applied. The point of summarizing the comparative analysis is to suggest that the Table 2–2 cost estimates based on use of platforms are probably a maximum in the sense that an alternative technique is available which is likely to cost less.

Water depth and climate-oceanographic conditions tend to be the primary factors influencing the investment requirement, but there are several other elements which can influence cost. Some of these factors are distance of the platform from shore, nature of the effluent, depth of the hydrocarbon reservoir beneath the ocean floor, size of the area over which the deposit is distributed, and location of the operation relative to the nearest equipment and support facility.

If natural gas happens to be the output, the distance between the platform and the shore becomes a major variable influencing the total investment requirement. Natural gas can be safely and commercially evacuated from the production site only by pipeline. But seabed pipeline costs from two to four and a half times more than comparable onshore pipeline [22]. Water depth also influences the cost of the pipeline. For example, in 1972 a typical 24-inch diameter pipeline installed in the Gulf of Mexico cost $225,000 per mile for a 0- to 30-meter (0- to 98-foot) depth; $275,000 per mile for a 30- to 60-meter (98- to 197-foot) depth; and $400,000 per mile for a 90- to 105-meter (294- to 344-foot) depth. A 30-inch diameter pipeline cost $330,000 per mile in 0 to 30 meters of water and $475,000 per mile at water depths ranging from 30 to 60 meters [23].

The length of the drill hole beneath the surface of the ocean floor also affects the level of initial investment. Length is used instead of depth because the well may be set at a slant, thereby requiring a greater drilling length than the actual vertical distance. Most wildcat wells are dry—they do not strike oil or gas—so they are simply plugged with cement and do not involve further expenditure. To give some idea of exploratory drilling cost relative to the length of the hole, in the Gulf of Mexico in the early 1970s, the first 1803 meters (5916 feet) of a dry hole cost $100,000 and a 3050-meter (10,007-foot) length cost about $170,000.

The initial investment requirement can also be affected by the nature of the effluent. Natural gas has already been noted as a particular case in which the effluent, because of its relatively volatile nature, requires special handling and involves high costs. Aside from natural gas, the nature of the effluent pertains primarily to oil. In some cases the oil may have a particularly high viscosity or it may contain large amounts of sand or water thereby requiring special production or handling facilities [24].

The cost estimates from Tables 2–1 and 2–2 can be combined to indicate the total investment for equal capacity offshore complexes in various water-depth and climate-oceanographic conditions. Because of the appreciable span of time between the first expenditure and the first barrel of production, it is necessary to convert the disparate current values in Tables 2–1 and 2–2 into standardized future values [25]. Table 2–3 provides cost per barrel of output and future value of investment calculated at a 12 percent rate of interest for the number of years indicated beside each entry [26]. For the gestation period, it is assumed that exploration requires two years and that expenditures proceed at a constant rate over this period. It is further assumed that expenditure for the construction and placement of

Table 2–3. Future Value of Total Investment Expenditures in Exploration and Production

	Climate					
	Mild		*Moderate*		*Severe*	
Water Depth (Meters)	*Value ($\$10^6$)*	*Years*	*Value ($\$10^6$)*	*Years*	*Value ($\$10^6$)*	*Years*
200	117.60	4	141.83	5	401.36	6
500	345.65	5	414.19	6	—	—
1000	606.58	6	724.43	7	1,574.64	8

development and production facilities is made at a constant rate beginning in the third year and continuing for the balance of the period indicated [27].

The figures in Table 2–3 indicate that production of oil from water depths of 200 meters and more involves a much larger investment than is required for land-based production, and a substantially higher investment relative to shallow-depth offshore operations. In view of the very large investment requirement and lengthy gestation period, the question arises whether large-scale exploitation at water depths greater than 200 meters is a reasonable expectation. The response would depend on the disposition of several important variables, including future oil prices, the size of individual petroleum deposits, the cost of capital and technological change.

The first important variable is the international price of oil in the decade of the 1980s. The viability of offshore production beyond water depths of 200 meters will depend partially on whether there is a downward change in the 1974 international oil price relative to estimated 1974 offshore production costs.

Offshore construction costs approximately doubled between 1950 and 1965, and since 1970 the rate of cost increase has accelerated [28]. It remains difficult to judge, however, whether cost per unit of output, for given production conditions, has been increasing because technical change so permeates the entire offshore oil production process. For example, a National Petroleum Council study found that during the period 1950 through 1960 total cost reductions of \$1.00 per barrel were effected by technical improvements in drilling, production, corrosion control and well spacing [29]. In absolute terms, offshore production costs can be expected to reflect the general inflation rate. Production costs are a composite of several diverse variables, e.g., steel, skill and technology prices—prices which originate in several different industries involving competing companies around the world. This implies that input costs are generally subject to competitive forces and are therefore not likely to experience abrupt or unpredictable changes. Thus, while forecasts of offshore production costs are not definitive, policy-makers can still use recent past and present cost behavior as a reasonable basis for decision-making about future investment activities.

As opposed to forecasting oil production costs, the prediction of an oil price is much more difficult. Since 1972 the international price of oil has demonstrated an appreciable upward volatility. The underlying potential for rapid and essentially unpredictable price changes is large because the international price of oil is virtually unrelated to the cost of production. This is a partial reason for the fact that the international price of oil can be and is an administered price. Another fundamental element which enables abrupt and large oil price changes is the fact that the quantity of oil consumed during the first year or two after a price increase is not likely to decline in proportion to the price increase. Finally, the world's major oil exporters have very large trade surpluses and certain foreknowledge that large export surpluses can be continued into the future so long as they cooperate with one another. Therefore, on economic grounds, the major oil exporters can be relatively insensitive to the results of price increases, assuming their basic cartel arrangement is maintained.

Direction of the international price of petroleum was largely taken over by the Organization of Petroleum Exporting Countries (OPEC) in 1973. The price of oil in the 1980s may or may not be determined by OPEC; but if it is predicted that OPEC will not be able to set the price in the 1980s and that the price will be lower in 1974 terms, it becomes necessary to provide an explanation.

Aside from events of a political nature which might cause the price to fall, an explanation based on economic grounds would have to involve prediction of a long-run decline in demand, a long-run increase in supply, or some combination of the two, together with simultaneous acquiescence by OPEC to allow these changes on the supply side to occur. That the requisite market changes will occur is conceivable as one combination of several possible economic events, and it is conceivable that OPEC will alter its policy because of market pressure, e.g., a situation in which oil importers reduce their imports so much that individual oil exporters violate their agreements and start cutting prices, or one in which new oil exporters enter the market and underprice a sufficiently large volume of petroleum to force OPEC to reduce its prices.

Indeed, analysts have made oil price forecasts, but there is little consensus. Using what might be called an intuitive or judgmental approach, M.A. Adelman argues that because the price of oil is "10 to 20 times cost . . . supply and demand have nothing to do with the world price of oil: only the strength of the cartel [OPEC] matters" [30]. Adelman asserts that if international oil companies were to relinquish their production role, the cartel would rapidly erode and break under the strain of tremendous competitive pressures caused by the large differential between production cost and selling price, resulting in a precipitous drop in the world price of crude oil. Employing a similar informal approach but reaching an opposite conclusion, James W. McKie voices the opinion that the OPEC cartel may be able to persist because the long-run ability to increase petroleum production from non-OPEC sources is not likely to outpace increases in long-run demand, even allowing for increased oil prices [31].

Several writers have constructed formal analyses of the international petroleum market whereby, using either simulation or optimizing techniques, they forecast the long-run price path of petroleum [32]. While all of their analyses are not directly comparable and of course differ in assumptions used and results obtained, most indicate that the international price of oil will not rise substantially in real terms above the 1974 price level during the period 1975–1985. Dietrich Fisher *et al.*, in their excellent critical review of the major formal forecasts, note that the results of most of the models suggest that $5 to $6 per barrel is a floor price, virtually regardless of OPEC actions, and that "the most likely price range becomes $7–$10 [per barrel] with general agreement on an expected value of about $8.50" [33].

As a third approach, the implications of certain theoretical economic models have been examined based on the premise that the models reflect the salient characteristics of OPEC and its economic milieu. As opposed to the preceding approaches, the theoretical models are not intended as price forecasts per se. One such theoretical treatment interprets OPEC policies as those of an optimizing monopolist [34]. Under this interpretation the cartel may rationally seek to maximize the present value of profits if the response from competing substitutes and other nonmember crude oil producers is restricted to the distant future. Thus the price of oil may not decline for some time, and then the decline may be slow. Another theoretical approach describes the world petroleum industry as a duopoly arrangement in which one part of the duopoly consists of an OPEC-like cartel which dominates price determination by regulating its output in a specific manner. The second part of the duopoly consists of

Table 2–4. Earnings from Offshore Petroleum Production

				Climate		
				Mild		
Water Depth (Meters)	Price ($/bbl.)	Net Annual Earnings ($10^6)	Recoverable Reserves (10^6 bbl.)	Annual Capital Costs ($10^6)	Annual Net Revenue ($10^6)	Net Revenue ($/bbl.)
200	6.25	54.7	50	30.58	24.12	2.64
			100	19.76	34.94	8.83
	8.50	75.2	50	30.58	44.62	4.89
			100	19.76	55.44	6.08
	11.65	103.9	50	30.58	73.32	8.03
			100	19.76	84.14	9.22
500	6.25	54.7	50	89.87	Negative	Negative
			100	58.07	Negative*	Negative*
	8.50	75.2	50	89.87	Negative	Negative
			100	58.07	17.13	1.88
	11.65	103.9	50	89.87	14.03	1.54
			100	58.07	45.83	5.02
1000	6.25	54.7	50	157.71	Negative	Negative
			100	101.93	Negative	Negative
	8.50	75.2	50	157.71	Negative	Negative
			100	101.93	Negative	Negative
	11.65	103.9	50	157.71	Negative	Negative
			100	101.93	1.97	.22

*If the interest rate were reduced from 12 percent to 8 percent, all estimates of net revenue increase appropriately and the negative entries shown with an asterisk become positive.

all other sellers of oil, who act in a competitive way but in response to the price set by the dominant producer. Given the specification of the "dominant extractor" model, the price of oil increases over time until the stocks of competitive sellers are exhausted, at which time the price increase assumes a slower growth rate [35].

Finally, the International Petroleum Institute (IPI) projects a price of $16.12 per barrel in 1985. After 1985 competition from petroleum substitutes will, in the judgment of the IPI, attain a volume sufficiently large to cause the price of oil to remain relatively constant [36].

No position on the future price of crude oil is taken here except to

Table 2–4. continued

Climate					
Moderate			*Severe*		
Annual Capital Costs ($10⁶)	*Annual Net Revenue ($10⁶)*	*Net Revenue ($/bbl.)*	*Annual Capital Costs ($10⁶)*	*Annual Net Revenue ($10⁶)*	*Net Revenue ($/bbl.)*
36.88	17.82	1.95	104.35	Negative	Negative
23.83	30.87	3.38	67.43	Negative	Negative
36.88	38.32	4 20	104.35	Negative	Negative
23.83	51.37	5.63	67.43	7.77	.85
36.88	67.02	7.34	104.35	Negative	Negative
23.83	80.07	8.77	67.43	36.47	3.99
107.69	Negative	Negative	Not		
69.58	Negative*	Negative*	Available		
107.69	Negative	Negative			
69.58	5.62	.62			
107.69	Negative	Negative			
69.58	34.32	3.76			
188.35	Negative	Negative	409.41	Negative	Negative
121.70	Negative	Negative	264.54	Negative	Negative
188.35	Negative	Negative	409.41	Negative	Negative
121.70	Negative	Negative	264.54	Negative	Negative
188.35	Negative	Negative	409.41	Negative	Negative
121.70	Negative*	Negative*	264.54	Negative	Negative

say that the available material, empirical and theoretical, implies that a long-run real price approximating or something less than the price prevailing in 1974 is quite plausible. To accommodate different oil price forecasts, three alternative prices are included in Table 2–4. The intention of this table is to illustrate how the principal variables of price, deposit size, water depth, climate-oceanographic conditions and capital cost would combine to affect the viability of offshore operations.

The first price listed in Table 2–4 is taken from an econometric model by Michael Kennedy [37]. Kennedy asserts that the price for crude oil is not likely to be sustainable over time because the larger producers in OPEC "would have the problem of allocating reduced production levels among the individual countries" [38]. For example, Kennedy's analysis predicts that a $7 export tax would lead to an export volume reduction by 1980 equal to 125 percent of Saudi Arabia's 1973 peak output [39]. Kennedy's analysis indicates that an export tax of $5.25 is the maximum price in 1980 that will not yield a production surplus sufficiently large to break the cartel arrangement. To arrive at the $6.25 price in Table 2–4, one dollar is added to the export tax to allow for production costs. The second price in Table 2–4 is the "general agreement" price from Fischer's survey of formal oil price models. The third price is the posted price of light Arabian crude in January 1974 [40]. It is noted that if the International Petroleum Institute judgment concerning prices and volume of petroleum substitutes is correct, then the $11.65 price may be interpreted as essentially a long-run ceiling price.

For a given water depth and climate combination, six net revenues are calculated based on combinations of the three alternative prices and two different amounts of recoverable reserves. The annual capital cost is the amount necessary to recover the investment indicated in Table 2–2, while the unpaid balance simultaneously draws 12 percent interest for the period of time it takes to exhaust the deposit. Annual net revenue is the price per barrel multiplied by annual output (9.125 million barrels) less an operating cost per barrel set at $5000 per well per month, equivalent to $.26 per barrel [41]. Net revenue per barrel is calculated by taking the difference between annual net revenue and annual payments on the investment and dividing by annual output. Secondary recovery efforts are not included in the calcuations. If the annual net revenue is less than the annual amount necessary to amortize the investment plus accrued interest on the unpaid balance, the undertaking is not feasible. In Table 2–4 this is indicated by a "negative" entry under revenue per barrel.

The estimates in Table 2–4 indicate that a price of $6.25 per barrel is adequate to yield a positive return in 200 meters of water in mild and moderate climate conditions, but not in severe conditions, or at a 500-meter water depth under any climate conditions. The minimum viable price at the 500-meter depth is $8.50, and then only for the larger reserve level; however, if the interest rate is lowered from 12 percent to 8 percent, all net revenues increase and the negative entries bearing asterisks become positive. Production is not financially viable at the 1000-meter depth for the two lower prices, and it is viable at the $11.65 level only in mild conditions.

The calculations in Table 2–4, based as they are on a few simple assumptions, cannot indicate whether offshore exploitation beyond 200 meters will in fact be financially viable. They do illustrate the importance of such factors as price, the amount of recoverable reserves and the interest rate and suggest that exploitation could be viable for a limited portion of the seabed lying under more than 200 meters of water. The range of viability with regard to the amount of reserves, price and the depth-climate-oceanographic variables would conceivably increase if a different technology, such as use of the subsea completion system, were employed.

As a final point on the topic of commercial viability, it is worthwhile to consider briefly the effect of the amount of recoverable reserves on viability and how this factor relates to offshore petroleum potential. It may be recalled that in Chapter 1 thirty developing economies were listed as having a low petroleum potential (one billion to .01 billion barrels) and that offshore potential referred to a geological estimate as opposed to an economic estimate. Given the likelihood that the amount of economically recoverable oil is less than the amount actually existing and the critical importance of the amount of recoverable oil present in a specific area, the probability of offshore oil production for the thirty low-potential countries appears to be very small.

It will be recalled that in Weeks' judgment, 25 percent (or less) of offshore petroleum potential will be commercially viable; in addition, Weeks states that "operations under 330 meters or more of water may be so costly that only giant fields [500 million barrels] will perhaps be profitably exploited" [42]. Furthermore, when it is considered that the highest of the low-petroleum potential LDCs may have only 2.5 to 250 million barrels of economically recoverable petroleum, and to be recoverable it must exist in relatively large collections in locations which do not lie at excessive water depths, the chances that the low-potential group will exploit grow increasingly slim. It is not certain that all of the thirty countries listed in Table 1–9 will not eventually produce any hydrocarbons from their offshore areas, but the sum effect of the limiting factors suggests that a substantial number of LDC offshore areas will not yield commercial deposits.

While no claim is made that none of the thirty low-potential LDCs will ever produce offshore, in further discussion this number will be added to the group of twenty-five no-potential states to represent the approximate number of developing nations which will never develop substantial amounts of offshore hydrocarbon resources.

This chapter has reveiwed the technical aspects of offshore oil exploitation to ascertain the approximate outer technical limits on recovery of oil from subsea deposits. It is evident from the review of technical capability that it is technically feasible to produce petroleum from the continental margins under a broad range of water-depth, distance-from-shore and climate-oceanographic conditions. In the second part of this chapter, exploration and production costs were examined under different conditions and costs were compared to three alternative prices. In an illustrative exercise it was found that offshore oil exploitation could be commercially feasible for a segment of the world's continental margins beyond 200 meters water depth under certain circumstances. The existence of a technical ability to exploit offshore oil under almost any conditions and the existence of a price for oil sufficiently high to cover the cost of operations under most conditions suggest that those areas with high potential for offshore oil will realize substantial, new developments in offshore activity.

NOTES TO CHAPTER 2

1. Kash, *et al.*, *Energy under the Oceans*, pp. 28–31.

2. Between 1949 and 1974 the mobile drilling rig fleet expanded to approximately 300 rigs which are operating in thirteen major areas of the globe, including the coastal waters of southeast Asia, Africa, South America, Japan and Australia, and in the North Sea, the Mediterranean and the Gulf of Mexico. ("Mobile Units," *Offshore* 34 [July 1974]:99–114). It has been projected that there will be 400 units in operation by 1984. (P.J.M. Verschure, "How Long Will Offshore Drilling Be American-Dominated?" *Oil and Gas Journal* 71 (December 31, 1973):138.

3. This operation exemplified two significant technological advances. First, it was carried out using the dynamically positioned ship. The ship was maintained on an exact location by means of eleven sideways-acting propellers, two main screws and an extremely sophisticated control system directed by two computers operating in tandem. The ship stayed on one point for 32 days withstanding 65-mile-per-hour winds and ten-foot waves. ("Off Borneo: SEDCO 445 Drills with Dynamic Positioning," *World Oil* 176 [February 1, 1973]:31.) Second, reentry of the drill hole was achieved at a 396-meter depth. The demonstration of reentry capability is significant because a part of the drilling operation involves routine removal of the drill string to replace the worn drill bit. Reentry is therefore essential if a drilling depth great enough to reach a deposit is to be achieved.

4. National Petroleum Council, *Ocean Petroleum Resources*, p. 5.

5. A. Craven-Walker, "How BP Will Equip the North Sea's Forties Field," *World Oil* 176 (May 1973):55.

6. "Unique Platform Designed for 650-Foot Water Depth," *World Oil* 176 (February 15, 1973):133.

7. Kash *et al.*, *Energy under the Oceans*, p. 50.

8. Nicolò Alparone, "Gravity Platform Designed To Reduce Offshore Costs," *World Oil* 176 (February 15, 1973):90.

9. Kash *et al.*, *Energy under the Oceans*, p. 50.

10. The "SEAL" (developed by the Subsea Equipment Association Ltd., an international engineering consortium) undersea production system is a specific

example of the latter approach. The SEAL is an automated system with some remote control features. It uses relatively less expensive conventional equipment and can be serviced by personnel transferred to it by a diving bell apparatus. The SEAL has been tested in 457 meters (1500 feet) of water and has a potential operating depth of 914 meters (3000 feet). ("Subsea Production System Will Be Used in 1975," *Offshore* 34 [May 5, 1974]:44–45).

11. E.L. Dougherty *et al.*, "A Comparative Study of the Economics of a Subsea and a Platform Oil Producing System," paper presented at "Oceanology International 72," conference in Brighton, England, 19–24 March 1972, p. 400.

12. National Petroleum Council, *Petroleum Resources*, p. 93.

13. Kash *et al.*, *Energy under the Oceans*, p. 52.

14. H. LeLeuch and J. Masseron, "Econcomic Aspects of Offshore Hydrocarbon Exploration and Production," *Ocean Management* 1 (December 1973): 296–97.

15. National Petroleum Council, *Petroleum Resources*, p. 51.

16. LeLeuch and Masseron, "Economic Aspects," pp. 296–97.

17. "North Sea Costs and Returns," *The Petroleum Economist* 42 (December 1975):458.

18. UN Committee on Peaceful Uses, *Economic Significance of Limits.*

19. Hollis D. Hedberg, "Why Explore the Deep Offshore?" in *Exploitation and Economics of the Petroleum Industry: New Ideas, New Methods, New Developments* (Houston: Gulf Publishing Company, 1967), p. 62; and UN Committee on Peaceful Uses, *Economic Significance of Limits.*

20. "A new-field wildcat is a well located on a structural feature or other type of trap which previously has not produced oil or gas." It generally involves the highest degree of risk of the various types of exploratory drilling. (Fred J. Wagner, "North American Drilling Activity in 1974," *Bulletin of the American Association of Petroleum Geologists* 59/8 [August 1975]:1280).

21. The 20 percent success ratio can be taken only as an approximate magnitude. In 1974, new-field wildcat success ratios were 25.7 percent in Canada, 11 percent in Mexico, and 14.24 percent in the United States (*Ibid.*, p. 1273). Tanzer indicates that a 20 percent success ratio was achieved in petroleum exploration in the United States for those cases in which the assistance of geological and geophysical exploration were employed in advance of drilling (Michael Tanzer, *The Political Economy of International Oil and the Underdeveloped Countries* [Boston: Beacon Press, 1969], p. 125).

There is little statistical evidence on which to base an assumption for a deep offshore success ratio because there has not yet been sufficient activity to provide a statistical data base. Furthermore, any reference to an offshore success rate tends to be misleading because of the way in which successes are defined for statistical purposes and procedural differences between onshore and offshore operations. If an onshore new-field wildcat makes a discovery, development and production activity can be routinely established on the same site. If an offshore new-field wildcat strikes oil, routinely it is plugged and several other exploratory or "delineation wells" are drilled to ascertain the dimensions of the deposit and thereby determine the optimum site for a multi-well development and production platform. Because a success is defined as a wildcat which actually produces petroleum, the offshore discoveries which are subsequently plugged are not, by definition, successes. An attempt is currently under way to formulate a definition which accommodates the operational necessities of offshore exploiration (Fred J. Wagner, telephone conservation, May 19, 1975); however, the issue might be resolved with the increasing use of subsea completion systems which allow production from the discovery well and thereby eliminate the need to determine an optimum platform site.

22. National Petroleum Council, *Petroleum Resources*, p. 94.

23. Kash *et al.*, *Energy under the Oceans*, p. 88.

24. For example, it is conceivable that a petroleum deposit containing oil of extraordinarily high viscosity might require injection of steam to transform the oil into a sufficiently fluid state to allow extraction. Such a process might not be feasible, or it might be prohibitively expensive to have a subsea steam-injection capability. A not uncommon problem is dealing with the natural gas which is produced along with the oil. After the gas is brought to the surface it becomes necessary either to pipe the gas to shore, to burn or "flare" the gas at the wellhead, or to reinject it into the formation. Frequently, piping ashore is too expensive and, in some cases, flaring is not allowed. Reinjection, however, is also very costly and involves substantial technical problems.

25. The future value is the sum of annual expenditures with interest compounded on each annual amount from the time of expenditure to the end of the gestation period. Base annual expenditures are the total present value divided by the number of years in the gestation period.

26. Specific data are not available which indicate discovery and development times. However some information is provided by Henry L. Berryhill, Jr., who states that "the time required to produce after initial discoverty in frontier [offshore] areas will likely remain two to five years for those areas that are in water depths of less than 300 feet (91 meters) and will be in the range of seven to fourteen years for those that are in water more than 400 feet (121 meters) deep" (U.S. Department of Interior, United States Geological Survey, *The Worldwide Search for Petroleum Offshore: A Status Report for the Quarter Century 1947–72*, by Henry L. Berryhill, Jr., Geological Survey Circular 694 [Washington, D.C., 1974], p. 15). Berryhill estimates that discovery and development time for a giant offshore petroleum field with "offshore geology not related to [the] onshore producing area" is 10.5 years (*Ibid.*, p. 12). The not-closely-related-geology factor is more likely to be relevant as exploitation extends into deeper water. It is further indicated that sixteen to twenty years would be required to achieve a production level of 250,000 barrels per day from a North Sea oil field (and a total capital cost, in 1972 present value terms, of $758 to $1004 million (*Ibid.*, p. 14).

27. To simplify the estimation procedure, a simple "client country-producing company" arrangement is assumed. A company (or some number of companies, domestic or foreign) sells exploration and construction services to the client country. The client country assumes the risks and collects any residual from the selling price after deduction of costs. While this assumption does not necessarily reflect typical company-client country relationships, the simplification does not seriously alter the results. More typically the company or companies assume

some or perhaps all of the risk in return for a presumably increased share of the residual. The aggregate result is the same because, if risk is distributed randomly, the aggregate return will remain the same regardless of whether the countries assume all risks and rewards are shared between clients and producers.

28. National Petroleum Council, *Petroleum Resources*, p. 52.

29. *Ibid.*

30. M.A. Adelman, "Is the Oil Shortage Real? Oil Companies as OPEC Tax-Collectors," *Foreign Policy*, No. 9 (Winter 1972–73), p. 94. A similar argument is expressed in Adelman's "Politics, Economics, and World Oil," *American Economic Review: Papers and Proceedings of the Eighty-sixth Annual Meeting of the American Economic Association* 64 (May 1974):58–67. See also J.E. Atkins, "This Time the Wolf Is Here," *Foreign Affairs* 51 (April 1973):462–90.

31. James W. McKie, "The Political Economy of World Petroleum," *American Economic Review: Papers and Proceedings of the Eighty-sixth Annual Meeting of the American Economic Association* 64 (May 1974):51–57.

32. Dietrich Fischer, Dermot Gately and John F. Kyle, "The Prospects for OPEC: A Critical Survey of Models of the World Oil Market," *Journal of Development Economics* 2 (December 1975):363–86. Two other relevant articles are included in this issue: Charles Blitzer, Alex Meeraus and Ardy Stoutjesdijk, "A Dynamic Model of OPEC Trade and Production," 2:319–35; and Basil A. Kalymon, "Economic Incentives in OPEC Oil Pricing Policy," 2:337–62.

33. Fischer, Gately and Kyle, "The Prospects for OPEC," p. 375.

34. John F. Kyle and Warren E. Moskowitz, "The Economics of the OPEC Cartel: A Theoretical Discussion," New York University no. 74–20, October 1974. (Mimeographed).

35. Stephen W. Salant, "Exhaustible Resources and Industrial Structure: A Nash-Cournot Approach to the World Oil Market." *Journal of Political Economics* (forthcoming).

36. International Petroleum Institute, *1985: A Dual Impact Study of Our World in 1985 before and after the Energy Crisis* (New York: M. Barrows & Co. and International Petroleum Institute, 1975).

37. Michael Kennedy, "An Economic Model of the World Oil Market," *The Bell Journal of Economics and Management Science* 5 (Autumn 1974): 540–77.

38. *Ibid.*, p. 540.

39. *Ibid.*, p. 567.

40. *Petroleum Intelligence Weekly*, 14 January 1974, p. 6.

41. Kash *et al.*, *Energy under the Oceans*, p. 86.

42. UN Committee on Peaceful Uses, *Economic Significance of Limits*, p. 17.

Chapter 3

The Economic Interaction of Developing Economies with the Offshore Petroleum Sector

There can be little doubt that developing nations which border on the oceans will ultimately (if they have not already) extend national authority over their adjacent continental margins and the petroleum resources trapped therein. Eventually several developing economies will by some means initiate offshore petroleum production; but it remains an open question what role they will play in the hydrocarbon exploitation operations themselves. To help illuminate this question, technical, financial and other aspects relating to developing economy participation are considered and alternative participatory arrangements are outlined. The last section of the chapter is an examination of the basic kinds of positive and negative economic effects on developing economies (those with offshore production and those without offshore production) brought about by the initiation of offshore production.

THE ROLE OF THE DEVELOPING NATION IN OFFSHORE OPERATIONS

Over the last twenty-five years, less developed economies, whether major or minor petroleum producers, have tended to assume an increasing degree of control in the development of their petroleum resources, and the momentum of this movement continues to increase. The fundamental question in the development of the areas off the coasts of developing nations, with particular reference to the geological formations which are not extensions of onshore formations, is whether the momentum of increasing LDC control will be maintained and extended into their offshore areas or if it will be blunted by the technical challenge and high-risk investment inherent in offshore exploitation.

For several decades, the argument has been made that the risk and the cost of petroleum exploration were so great that only international oil companies could successfully develop petroleum. Their losses could be offset by gains in other areas. It was further argued that only the international oil companies could attract and organize the necessary technical and management expertise to develop petroleum resources successfully. Finally, it was argued that only the large integrated companies could marshall the large amounts of risk capital required for petroleum resource development [1].

The claims of international oil company advantage over LDC national oil companies may have had some validity at one time, but the trend toward increasing LDC national control and direct involvement in development of their petroleum resources suggests that the strengths of the international companies are not uniformly great across countries and that these strengths have steadily eroded over time, as has the dominant position of international companies in the determination of LDC petroleum resource development [2].

For a number of reasons, developing nations have become more motivated and better able to assume a portion or all of the functions served by international companies. Motivation has increased with increasing nationalism and decreasing incidence of small (non-popular) ruling elites and monarchical governments. Certainly the quadrupling of crude oil prices in 1973–1974 has served to weld petroleum development policies firmly into each developing economy's basic set of major policy concerns [3].

Offshore operations have tended to retain a larger degree of foreign oil company involvement relative to onshore operations, but some developing economies have ventured into the oceans. Recently Petrobras, the state oil monopoly of Brazil, initiated searches along the coast off Rio de Janeiro and succeeded in discovering one field of approximately 900 million barrels, which more than doubles Brazilian oil reserves [4]. India has followed a relatively independent policy in exploring its Bombay High offshore area, but progress has been slow, and some American technical input has been called on to help deal with technical problems. The People's Republic of China has demonstrated a considerable ability to exploit offshore, including fabrication of major pieces of equipment. Pemex, the Mexican oil monopoly, has also undertaken offshore operations.

The very high price-cost ratio in the 1960s and 1970s provided a strong underlying element in the push for greater LDC participation.

Now it appears that the higher range of the price-cost ratio for deep-water offshore production could overlap the lower range of the price-cost ratio prevailing in the 1960s. That price-cost ratio varied between approximately 20:1 (for the Middle East) and 7:1 (for Venezuela). In the 1980s, deep water offshore crude might be produced at a cost as low as $2 a barrel and sell for as much as $10 to $16 per barrel, for a ratio range between 5:1 and 8:1. Again, it is emphasized that large variations in production costs can be anticipated, but the rough evidence suggests that the price-cost ratio will continue to provide an appreciable incentive for participation in production.

Recognizing increased level of LDC involvement in development of their own petroleum resources, there is also reason to believe that developing nations may be major participants in development of their offshore petroleum resources. But there are also reasons why it may not be feasible for less developed economies to pursue a totally independent path, even allowing for purchases of equipment from the industrial countries. First, with reference to events in the last twenty-five years, the assertion of LDC control over petroleum production and development was facilitated by the fact that the exploration, development and production phases had already been completed by the international oil companies. The prevailing exploration and production conditions (relative to the deep-water offshore conditions which are now contemplated) were substantially less difficult, and knowledge of the necessary technology in the pre-1970 period has become relatively well known and widespread.

The preceding section on technology provides evidence which suggests exploitation of offshore petroleum will tend to reintroduce the technology-intensive, high-risk, high-cost and difficult production aspects which were prevalent in development of onshore petroleum resources in the early stages of the postwar era. Thus, it is not clear that the services of international oil companies and smaller, specialized firms will be dispensable for all or most developing countries in their offshore areas.

The unique characteristic of international oil company participation is that they have the technical and managerial capacity and the experience which enable them to provide overall coordination of development programs. Deep-water offshore operations, as a new and different kind of undertaking, will create a heavy demand for the intangible but vital ingredients of technical innovation and problem-solving ability which international oil companies have the ability to deliver. As an alternative to comprehensive services provided by international oil companies, some less developed economies may choose to purchase the services of specialist firms. In fact, the international companies routinely subcontract certain aspects of offshore work to specialists because of the extraordinarily complex nature of the work. Even the technically sophisticated nations bordering on the North Sea rely heavily on foreign specialist firms such as (American) drilling companies to perform most of the offshore drilling operations [5]. Many other technical services are associated with offshore exploitation, such as well servicing, special exploration techniques and interpretation of data. The total absence of limited amounts of imported technical services and equipment (e.g., subsea completion systems) in a developing country offshore development program seems highly improbable.

There are other reasons to believe that the degree of developing economy involvement in deep-water offshore oil exploitation will not be as great as has been the case in onshore operations. One of the fundamental deciding issues is the expected rate of return on investment in offshore operations and relative ability of international firms vis-à-vis national firms to estimate and to influence (within limits) the margin of error related to the expected return rate. The expected rate of return calculation requires that the estimate of discounted income from production be weighed against the probability that there will not be a discovery of sufficient size to cover exploration, development and production costs plus a return on the basic investment [6].

Of course, there are several key risk estimates which must be included in the overall computation. First, there must be an estimate of the risk of discovering enough petroleum such that the minimum amount produced will be sufficient to yield a discounted income stream adequate to cover all costs. Next there are the risks of encountering delays and additional costs from technical problems once a discovery is made. For foreign companies, there is the risk of uncompensated nationalization or some form of excessive taxation. And finally there is the risk that the price of oil will change downward in the future.

The international oil firms remain strong in their ability to evaluate and, to a degree, to limit exploration-discovery risks because they possess or have ready access to the requisite technical skills, experience and the most advanced equipment. Because of their extensive experience, most international oil companies also tend to be better prepared to overcome technical problems which might be encountered in the process of development and establishing production. Finally, the international oil companies can diversify their risk portfolios by selecting a number of exploration sites within the region—

or around the world, if necessary. The diversification advantage also applies to technical risk and to nationalization risk, although the latter point is not an advantage over a national undertaking. To the extent that the less developed economy possesses a seabed area with extremely high petroleum potential, the risk factor related to the discovery phase may be so small that the advantage of the (presumably more expert) international company is insignificant. If the developing nation has rights to an area which provides an extremely large number of petroleum possibilities, the national firm also has an opportunity similar to that of the international firm to spread its exploration risks. Thus, in these two instances, the national firm might enjoy an advantage similar to that of the international firm.

The estimate of risk of a future downward price movement would seem to be the same for both a national and an international company; however, this is not necessarily the case. Deep-water offshore exploitation involves a much longer gestation period than an onshore undertaking, and the offshore operation has a greater potential for encountering technical problems which can prolong the gestation period. Therefore, the international firm may have some advantage in minimizing the downward price change risk by virtue of the probability that it can commence production earlier than a national firm for a given production situation. This difference, however, must be considered to be extremely minor.

In summary, if the problem is viewed as one of relative abilities to make the average realized return (the sum of the actual returns from several projects divided by the number of projects) approximate or exceed the expected return on investment in each project, then the international oil company can be expected to play a major role because it is well suited (1) to estimate the risk associated with the various aspects of deep-water offshore petroleum exploitation, (2) to minimize certain kinds of risks (dealing with technical problems or obtaining and evaluating geological data) and (3) to diversify risks among several different locations such that there is a greater chance that the average return realized from the several locations will approximate expected return.

The financing of investment in offshore exploitation is no less a key issue than the control of technical inputs, in that greater financial involvement by developing nations corresponds to a larger role. Thus a developing nation can be expected to exercise more control and, if successful, earn a larger return the greater is its share in the total investment. In approaching the financing issue, the less developed economy faces three basic alternatives if it intends to develop its offshore petroleum resources. First, it may be able to finance the undertaking from its own savings [7]. Second, the country may be able to borrow from foreign banks, raise funds on the Eurocurrency market, or acquire loans from other governments or perhaps from regional lending institutions. The International Bank for Reconstruction and Development (IBRD) is a potential source of investment funds, but to date this institution has not made commitments for petroleum development projects. It has been the practice of the IBRD not to participate in financing development of LDC petroleum resources in spite of the fact that the IBRD's primary function according to the Articles of Agreement is to promote private foreign investment, and petroleum exploitation has traditionally been a prime area of private foreign investment [8]. According to one report, however, this policy may be altered as the result of a recommentation by an IBRD study group [9]. Ostensibly, the IBRD would act as a lender of last resort and seek to attract private investment; however, the study recommendation has yet to be adopted as an IBRD policy. Even if this new financial source were to emerge, proposed offshore activity would still have to compete with land-based operations for IBRD investment funds. Nevertheless, this new source of financing could be quite significant for some developing economies. The third financing source is oil companies as a part of their comprehensive managerial-technical-financial package. These three alternatives are not mutually exclusive, although the international oil company option might be, relative to the other two, if a developing country's entire offshore area were so small that it could be readily encompassed by one company.

Because of the magnitude of the required investment, it is probable that the self-financing alternative would be available to either the wealthiest developing nations (e.g., perhaps some OPEC members) or "intermediate-level" developing economies (e.g., Brazil) which are willing to foresake some portion of their other investment programs in favor of investment in offshore development. The foreign-borrowing alternative has a somewhat broader application, although it is still limited to countries with an established international credit rating and, if the loan is sought directly by the state oil company, to LDCs with a petroleum industry which is considered competent to evaluate and deal with the technical aspects and which is creditworthy [10]. Foreign credit also might be available to those developing countries with a good credit rating and which are known to have outstanding offshore prospects or which can develop evidence of favorable offshore potential (perhaps with some self-financed preliminary exploration). Foreign governments could be willing to provide investment funds in selective cases, just as some OPEC countries have expressed a willingness to support the development efforts of a few countries. Regional development banks historically have not made funds available for petroleum development, but this practice may not persist indefinitely. The role of regional development institutions can be expected to remain limited, however, because of the large amounts of capital required relative to their capacity to provide funds. In sum, the poorer economies, the developing nations with weak international financial credit standings and the countries with less-than-excellent offshore oil prospects are unlikely to be able to finance offshore petroleum development with savings or with major flows of foreign credit.

The choice for these residual countries (and others which may have the financial capacity but for other reasons prefer not to become directly involved) is to turn to the international oil companies for financing. This alternative, unlike the other financing modes, involves a comprehensive arrangement whereby financing is implicit because it is provided in kind in the form of equipment, materials and technical services. Thus, the LDC which chooses the third financing alternative simultaneously selects the operational approach to development of its offshore petroleum, that is, substantial international oil company involvement.

ALTERNATIVE EXPLOITATION MODELS

The preceding discussion is essentially an attempt to indicate the direction in which the relative balance of participation might move, but it is also helpful to have some idea of the particular degree of participation a low income country might assume. If all developing countries were homogeneous with respect to their resource endowments, policy inclinations and economies, then a single exploitation model could be postulated to describe every country's involvement. Taken together, however, developing nations present a broad mix of these elements, thus it is realistic to expect a spectrum of basic arrangements to develop in which the individual country plays more or less of a role depending on its particular set of characteristics. Six basic exploitation models can be envisioned in order of decreasing developing country independence.

1. The model with the smallest foreign input is characterized by an approach such as that employed by the People's Republic of China: foreign input is restricted to specialized and key pieces of equipment. Foreign technicians play a minimum role, and foreign firms do not take responsibility for any substantial aspect of the

overall operation. There is no appreciable foreign financial participation. The Japanese approach might also be found near this end of the spectrum.

2. A state company performs most of the operations but lets service contracts to foreign firms for particular tasks. Again, there is no appreciable direct foreign investment. An example of this model might be found in the operations by Petrobras in Brazil, or perhaps to an even lesser degree by Pemex in Mexico.

3. Mixed operations in which both foreign firms and the national company operate fully integrated operations simultaneously. Foreign companies lease tracts for exploration and subsequent development and production if discoveries are made. A national company also conducts a fully integrated and independent operation purchasing required technical services and equipment from foreign firms. The agreements made by the Peruvian government in the early 1970s are an example of this model. In the Peruvian case and others, it is common that the host countries gain some advantage in exploration because foreign firms are required to relinquish one-half of each tract to the state at the end of the (seven-year) exploration period if oil is discovered [11].

4. Joint operations in which foreign firms perform under a basic service contract arrangment but local workers and technicians are methodically integrated into operations. Equipment and other fixtures may become the property of the host country. The effect for the company which provides the capital and services is that it operates on a comprehensive service contract while the developing country becomes a kind of partner, acquiring skills and output, although exploration risks are assumed by the foreign country. The company receives payment in the form of crude oil. The joint operation characterizes the arrangments made with foreign companies by Pertamina and Petromas, the state companies for Indonesia and Malaysia, respectively [12].

5. International oil companies dominate operations, providing virtually all technical skills and financial support and assuming all risks. They do not, however, control development policy nor do they necessarily have a domestic marketing monopoly. This model might be characterized by the arrangements in some sub-Saharan countries and in South and Southeast Asia.

6. International oil companies essentially are the oil sector; they largely control petroleum development policy and domestic marketing. The government is a relatively passive tax collector, although some of the taxes may be in kind. So far as it pertains to offshore production, there does not appear to be a present situation to which

this model would closely apply, although it is remotely conceivable that such a case could arise [13].

A developing country's present capability does not necessarily dictate which model will be applicable. With the movement by developing nations toward greater control of their petroleum resources and the fact that extensive offshore exploitation is not likely to occur before 1980, some countries may as a matter of conscious effort change their participatory posture from what it otherwise would be without an explicit policy to develop an offshore production capacity. If there is reason to believe that a developing country has an appreciable offshore hydrocarbon potential, but it lacks the immediate necessary technical and financial ingredients to allow substantial participation with foreign capital inputs, the country could conceivably develop such a capacity. It would require, however, that it select development of its petroleum exploitation capacity as an explicit policy from among the set of investment alternatives facing it. The country could then proceed to build a cadre of technicians and skilled workers and to assemble an organizational apparatus to function as the foreign company counterpart. It might also be necessary to consolidate or otherwise reorganize and update national laws dealing with petroleum exploitation and to establish a policy-making group on offshore resources.

If an even greater degree of participation were desired, it would also be necessary to develop investment sources over the preparation period either by shaping domestic programs to accommodate offshore investment or by cultivating foreign commercial and government investment sources, or some combination of the two. The point is the prospects for developing country participation should not be viewed as totally static; less developed nations have demonstrated in the past the ability to acquire technical and administrative expertise as well as policy acumen to a degree sufficient to enable them to enhance substantially their position vis-à-vis foreign companies. It is not unreasonable to expect that some countries will be able to shift along the participation spectrum if they choose now to build up a technical-administrative infrastructure, to consolidate their resource policy apparatus and to assemble investment sources.

It is difficult to make an overall assessment of the role of developing nations in offshore exploitation, but it would seem generally correct to say that the lower the level of a country's economic development, the smaller the expected role of that country in development of its offshore resources. Such an assessment is essentially an alternative way of saying that the poorer nations of the developing world, by definition, have the smallest technical-administrative capacity; they do not have an investable surplus of any significance and they are the least able to obtain the necessary investment resources from foreign sources. This leads to the general conclusion that the weight of the technical and financial aspects of offshore exploitation tend to tip the scales in favor of major oil company involvement for a large number of countries while the more advanced developing economies may adopt a more independent offshore petroleum development mode.

This conclusion should not be taken to mean that less developed economies cannot shift along the participation spectrum. If there is reason to believe that offshore petroleum resources are sufficient to make the advance effort worthwhile, some developing nations may be able to develop an offshore technical-administrative infrastructure, establish domestic and foreign credit sources, and combine these efforts in the context of their ultimate control over the resources.

As a closing statement on developing country participation in offshore petroleum development, it can be said that there are probably few, if any, major surprises in store relative to existing arrangements. The exploitation of a particular country's offshore area will probably follow the basic pattern of arrangements as they now exist. The pattern may be in transition for any given country and, if it is, it is probably moving toward a greater degree of national participation. The host of technical problems posed by offshore and particularly deep-water offshore exploitation and the extremely large financial requirements will tend to attenuate to varying degrees the rate of change toward greater national control, wherever the particular country lies in the six-point spectrum. It is possible, and even expected, that unforeseen technical problems or capital shortages will move some countries to allow more foreign involvement than they otherwise might have allowed. Others may find themselves particularly adept at constructing an effective technical-administrative infrastructure and assembling financing, and as a result may actually increase their degree of involvement in offshore operations beyond present expectations.

THE ECONOMIC IMPACT OF OFFSHORE HYDROCARBON PRODUCTION

The prospect of offshore hydrocarbon production is not a new event, nor is it unique in an economic sense. Offshore hydrocarbon production has been taking place in developing nations for many years and, in countries such as Indonesia, Nigeria and Gabon, it has become a major economic feature. Production of hydrocarbons from offshore sites *per se* is not unique in an economic sense relative to exploitation of any other natural resource. Nevertheless, the opportunity to produce hydrocarbons from offshore areas could be a significant factor for developing economies.

Developing nations with offshore hydrocarbon potential will not necessarily find the prospect to be a readily programmable economic factor. A country may have an offshore area with a high potential, but the process of precise location of commercial deposits can be both time-consuming and uncertain. In the North Sea, for example, about 200 holes were drilled in the ocean floor before the discovery of natural gas, and 400 wildcats were drilled before commercial oil was produced [14].

The existence of two countries in the same offshore potential category does not necessarily mean that the actual discovery and production of oil will have the same economic significance for both countries, even if the amounts of oil and production conditions are the same for each of them. The degree of economic impact is partially dependent on the size of the economy relative to the magnitude of offshore production. For example, if India were ultimately to discover 10 billion barrels of oil in its offshore area and Thailand were to discover the same amount, the potential economic impact on the much smaller Thai economy (*ceteris paribus*) would be expected to be greater. Similarly, the relative degree of strategic (although not necessarily economic) impact can be expected to be greater for a country which is largely dependent on imported oil as opposed to a country which is self-sufficient in oil. Thus, the discovery of oil off the coast of Chile or Korea would have a different significance for those two countries than would discovery of the same amount of oil under the offshore waters of Nigeria or Indonesia.

Production from deep-water, offshore areas may be expected to be negligible for at least ten years [15]. And for those countries whose offshore areas do finally yield commercial oil production, offshore production still cannot be expected to be a panacea for economic development problems or a guarantee that rapid economic development will occur. Problems such as unemployment, excessive reliance on the primary production sector, low literacy rates or inadequate nutritional levels can be expected to persist unless explicit policies to deal with these problems are implemented. Nevertheless, production of oil from offshore areas could induce several effects which are generally thought to be desirable.

There are three basic kinds of quantifiable effects which would be likely to occur as a result of offshore exploitation: the first is a net gain in foreign exchange as a result of petroleum exports, taxation and/or foreign exchange savings as a result of reduction in petroleum imports; second, there is additional net investment specifically for the purpose of increasing future income; and third, there is an increase in expenditure for local inputs, i.e., labor and non-labor inputs [16]. The magnitude of each of the three effects will be a function of each developing economy with respect to its domestic investment capacity, eligibility of domestic labor, supply capacity for materials, the particular agreement which is struck between the country and foreign company or companies, and existing petroleum production and petroleum consumption.

The preceding three measurable effects do not include all of the possible favorable effects which could result from offshore exploitation. It is correct to say that there are few activities which could have less of a direct involvement in a developing economy than the highly capital- and skill-intensive processes which make up offshore petroleum operations; there is, nevertheless, some direct employment effect and an appreciable potential for linkages, particularly forward linkages [17]. The potential for backward linkages, e.g., development and construction of equipment or production of materials (viz. mostly steel products) used in offshore operations is limited, but there is some opportuntiy for participation in providing repair and overhaul facilities. It is noteworthy that development of this kind of support facility is not necessarily restricted to location in the producing country and because of this could have a positive effect (identical in principle to the quantifiable effects from petroleum production listed above) on one or more other non-offshore producing developing economies.

The role of Singapore as a major servicing facility for offshore operations in Indonesia is a prime example of this particular aspect of backward linkage. The Singapore example also serves to point out that development of a support facility is a function of the availability of such factors as location, existence of a skilled or readily trainable labor force, entrepreneurial skills and a suitable harbor. The facts that one large support facility can service an entire region and that there are prerequisites for establishment of such a facility, further suggest that this kind of backward linkage is not likely to be widespread among less developed economies.

The potential for forward linkages from offshore petroleum production is more substantial. Developing nations which can muster the necessary complementary inputs can establish or participate in establishment of industries which use petroleum as inputs. This fact has certainly been assimilated by many petroleum-exporting countries in the last decade, as witnessed by their establishment, or plans to establish, refineries, fertilizer plants and petrochemical complexes. Of course, the quantifiable effects of increased foreign exchange earnings, increased investment and greater demand for domestic resources, noted earlier, would also be applicable for evaluation of forward and backward linkage developments.

There is also a potential economic advantage to be realized by offshore-petroleum producers in the form of increased technology transfer through the training of workers, managers and technicians. This may occur as nationals acquire skills, as foreigners are phased out and replaced by nationals, and as local workers from the offshore industry find or create opportunities elsewhere in the economy.

Aside from the incremental increase in net foreign exchange earnings, the advent of offshore oil production could affect a number of elements in the foreign sector. Simultaneous with the increased foreign exchange inflow, the debt-servicing burden (expressed as the interest and amortization payments to foreigners divided by export earnings per time period) would decline and thereby afford the opportunity to exercise a more flexible management over foreign borrowing. It is also conceivable that the increased foreign exchange inflow in the case of oil exports would allow the country to maintain a smaller foreign reserve balance and to use the difference in more directly productive ways [18]. Increased foreign exchange earnings from the new oil exports (or from decreased oil imports) could also be expected to ease adverse pressure on the exchange rate and present the opportunity to remove or at least ease foreign exchange controls and perhaps ameliorate some restraints on trade. Petroleum export earnings tend to be relatively stable and more predictable relative to most commodity exports. Developing nations have long argued that fluctuating export earning levels are deterimental to their economic welfare [19]. Thus, to the extent that this claim is valid and petroleum exports have the effect of smoothing out export fluctuations, the country would realize an additional benefit. On the domestic side, as the balance of payments becomes less of a problem, the pressure for deficit financing can also be expected to diminish, and, as a result, that part of inflation attributable to budget deficits can also be expected to diminish.

To the extent that transportation costs from offshore sites are less than transport costs for imported oil, the country will experience a savings in petroleum transportation costs. If services of foreign carriers are used, the savings will be in foreign exchange. If shipping services are domestically-owned, the shipping services could be sold to foreigners or, over time, the resources redirected to other uses.

In an era of increasing petroleum prices, a developing country which finds and produces petroleum from its offshore areas gains an important terms-of-trade hedge. In the first instance, the extent of this hedge would correspond to the degree to which offshore production displaces oil imports. As domestic production displaces a greater portion of imports, the adverse terms-of-trade effect from an increase in the price of foreign oil declines, reaching zero at that point where offshore production enables the country to be self-sufficient. To the extent there is an exportable surplus as a result of offshore production, the terms of trade actually improve with increases in the international price of petroleum. The existence of a terms-of-trade hedge, or "automatic" ability to improve terms of trade in the case of net exports provides obvious advantages for minimizing interruption in other export-financed development programs, and it presents an increased potential to accelerate the pace of economic growth in the event of future oil price increases.

Finally, there is an increasing probability, over time, that some petroleum exporting countries will reduce their exports to conserve their oil reserves for their own use. Any developing economy which can produce oil from its offshore area also gains a degree of insulation from the disruptive effects of such a reduction in traded petroleum. Indeed the effect could be to increase the price of the country's petroleum exports.

The advantages associated with new or increased petroleum production tend to be so impressive that potential disadvantages are sometimes overlooked. Nevertheless, there are a number of problems which have been associated with the introduction of offshore production. In the 1950s and 1960s, several economists developed the proposition that export-oriented commodity production did little to benefit the producing country [20]. Instead, it was argued, most of the benefits from primary commodity exports flowed to the industrial countries, whose companies provided the capital to produce the commodity and which served as the principal consumers of the exported commodities. Basing their argument on the economic model of a foreign-dominated sector using capital-intensive means to produce commodities for export, it was further argued that the existence of the commodity export sector inhibited industrialization and development in the rest of the economy. The argument was elaborated in the form of three subarguments: (1) the dual economy arguments, which emphasized the drain of scarce entrepreneurship, technical skills and capital to the commodity export sector; (2) the argument that fluctuating world commodity prices inhibit development and that, over time, commodity prices decline relative to import prices, thereby depressing the producing countries' terms of trade; and (3) the argument that nonrenewable resources, such as hydrocarbons, should be retained to contribute to future domestic development.

It is not intended here to review the various points and counter points in what amounts to a major debate among economists on the real effects of investment in development economies' commodity export sectors [21]. Nevertheless, it is noted that the basis for the dual economy and declining terms of trade-fluctuating price agreements, as they apply to oil, are dated and that the second argument is severely weakened by the events of 1973–1974. The conservation argument, if interpreted to mean an absolute prohibition on production, is economically unsound unless unusual circumstances prevail [22]. If governments of developing nations were sufficiently negligent not to exercise their sovereign and economic powers over offshore petroleum development, the dual economy problem could develop. This result, however, is not an inherent consequence of offshore petroleum exploitation *per se*.

In the same vein, other potential but perhaps less economically sophisticated pitfalls can be seen waiting for certain countries which find petroleum off their shores. For example, there is a possibility that an economic opportunity loss may occur if the new foreign exchange inflow from the export of petroleum is not properly managed (with due respect for hindsight). Two basic types of problems can be sketched. First, a sudden and large inflow may cause decision makers to initiate ill-conceived expenditures, such as investment in certain kinds of manufacturing, which cause severe labor displacement and increase unemployment. Second, a temporary surfeit of funds could allow the opportunity for a nonreversible transfer of funds abroad into foreign investments or bank accounts, or into nonproductive expenditures on imported luxury goods.

Ronald McKinnon has recently postulated that adverse economic results could occur as a consequence of a sudden increase in foreign exchange earnings [23]. He suggests that in the short-run, when an economy is in the process of adjusting to a substantial increase in capital inflow (a task which immediately presented itself to the members of OPEC starting in 1973), the demand for nontradable goods (goods and services produced for sale only in the domestic market) will be increased, causing wages to be bid up in that sector. The labor-intensive tradable sector will suffer adverse effects as resources are drawn into the expanding nontradables sector and the prices of tradables remain relatively constant because of a pegged exchange rate. Carrying his analysis a step further, McKinnon suggests that if government wages are increased as a means to distribute some of the capital inflow, it is possible that under certain circumstances wages in some other sectors will also rise and as a result production efficiency could fall, unemployment develop and total output decline [24].

The McKinnon analysis is significant because the adverse effects McKinnon describes could conceivably come about as a result of earnest government efforts to channel the gains from offshore production into the economy and to distribute the benefits of increased foreign exchange earnings. Declining employment and output are not inevitable results, however. For example, it is possible to pursue a policy of investment in human and physical capital simultaneously with wage increases. If wage increases were maintained on a par with increases in investment, the resultant growth in productivity would tend to maintain demand for labor while output levels increased.

As a last point, attention is drawn to the fact that if pollution hazards are not minimized, other sectors of the economy, such as tourism and fishing, could experience severe economic losses. Insofar as these sectors have a relatively high employment level, the potential for widespread economic damage could be substantial.

In spite of the possible disadvantages, there is little doubt that, on balance, exploitation of offshore petroleum can exert a strong positive economic effect. The degree of the positive effect, however, given the volume of petroleum production relative to the size and stage of economic development of the country, is largely dependent upon the policies which are adopted to exploit the advantage. In the same vein, serious negative economic effects are avoidable if an active and careful approach is adopted to make use of the opportunity afforded by offshore petroleum exploitation.

Favorable economic effects are not confined exclusively to developing nations which are able to exploit offshore. It is conceivable that a few non-offshore producing developing economies could realize some marginal benefit as a result of offshore petroleum production occuring in other countries. One possible advantage has already been mentioned: the chance that a country which does not have offshore production may have the opportunity to benefit from a backward linkage by becoming a regional overhaul and refitting center. Probably very few coastal developing countries with or without offshore production could engage in this kind of activity and, of course, the activity would be limited to states which at least touch the edge of the ocean. There is perhaps a greater probability of non-offshore producing coastal nations becoming involved in the forward linkage process by merit of some locational or labor force advantage. It is conceivable that neighboring developing economies, producers and non-producers, might combine their resources in joint operations as a means of attaining a sufficiently large-scale operation (e.g., a refinery or a fertilizer plant) while ensuring full utilization. But overall it is unlikely that very many non-offshore producing countries will become involved in the forward linkage process, and the chance that landlocked countries will become involved is virtually zero.

Other minor advantages which might accrue to non-offshore producers are: reduced transport costs because of development of closer supply sources and the possiblity that the certainty of future petroleum supplies may be increased for those countries which have particularly close ties with other developing nations which become offshore producers. Finally, as offshore oil production becomes established, there would be fewer candidates for multilateral and bilateral concessional assistance. As a result, relatively more concessional assistance could be allocated to the non-offshore oil countries. This last point serves to stress the meager prospects for those countries without offshore production potential. The increased concessional assistance factor stands out because it is the only possibility which is relatively certain and which could be widespread among non-offshore oil nations.

There are also some potential disadvantages to the non-offshore oil countries as a result of offshore oil production by their neighbors. The greatest single loss to non-offshore producers is that a portion of the benefits accruing to states with offshore oil represents an opportunity loss to all other states. Extension of national economic jurisdiction into the international area irrevocably prevents other states from ever directly benefiting from exploitation of the nationalized area. Thus, part of the benefits which accrue to countries from exploitation of oil in their newly acquired offshore areas are losses to other states.

The dual economy problem was mentioned earlier in the context of possible disadvantages to countries which establish offshore production. The problem, simply stated, involves a wage and return differential-induced flow of capital, skilled labor and entrepreneurs into the offshore petroleum sector to the detriment of the rest of the economy. Developing nations which experience the creation and expansion of an offshore industry are likely to be very cognizant of the potential dual economy problem and to take appropriate precautions. Nearby economies which do not have an offshore sector or some strongly dynamic sector may not be as sensitive to the problem however, and, if labor and capital are free to migrate across national borders, neighboring economies may suffer a loss of these scarce resources.

While the dual economy phenomenon is a potential disadvantage to non-offshore petroleum countries, the real effect of the dual economy phenomenon, if it occurs, is not readily predictable. First, the amount of factor movement, assuming no explicit measures to constrain it, is limited because the amount of labor used in offshore production is small in absolute numbers and there are natural barriers to labor migration. Furthermore, if capital can move to the foreign offshore oil industry, presumably it can also move out of the economy to other foreign destinations where returns are higher. Therefore the addition of a nearby offshore oil sector may not constitute a substantial addition to aggregated capital outflow incentives. Finally, it can be argued that the movement of capital and labor to a neighboring offshore petroleum industry could be economically beneficial to the country of origin. Such migration could decrease unemployment in the country of origin, upgrade labor skills, increase foreign exchange earnings from remitted wages, and allow capital the opportunity to earn a higher return.

Neighboring coastal states which do not have an oil potential in their margins still are subject to a pollution hazard from offshore production in nearby areas, but there is no way to predict, in the absence of specific circumstances, whether or not adjacent or nearby states run a smaller or greater risk from offshore pollution. Oil escaping from a production site will not necessarily result in damage to nearby beaches or fisheries because prevailing winds and water currents can localize the spill or keep it away from even a nearby area. Conversely, wind and currents can also drive a spill into distant areas. There is evidence, however, that the harmful components of crude oil evaporate are dissipated within a few days; therefore, the more time taken by a spill to reach an economic area the smaller the amount of damage that is likely to occur [25].

It is much more difficult to generalize about the net economic impact on non-offshore LDCs as a result of offshore exploitation in other LDCs. For a limited few non-offshore oil producers, there may be a limited positive impact because of special circumstances. The opportunities for backward or forward linkages in second countries is likely to be extremely small. Perhaps the strongest positive aspect is the increased potential for receiving increased allocations of concessional assistance. Relative to their more fortunate neighbors, the noncoastal states are confronted by a major negative aspect in that they lose access forever to resources which formerly had been in an international area and which sometime in the future they might have been able to exploit.

It has been the purpose of this chapter to examine the probable role of less developed economies exploitation of their offshore petroleum resources. Recognizing the trend toward increasing developing economy involvement and control of their petroleum resources, the extreme technical complexity and the high-risk, high-investment nature of offshore exploitation, the conclusion was reached that less developed economies will tend to remain dependent on industrialized countries for technical and capital inputs to a substantial degree. On the other hand, if those developing countries with more promising offshore prospects take steps now to develop a technical and administrative infrastructure and perhaps acquire some financial backing, their relative involvement could be substantially greater than an assessment of present capabilities would indicate. The control of offshore pretroleum development policy is likely to rest with the developing nations in spite of the substantial technical and financial elements of offshore operations. In view of the large differences among countries in their relative degrees of development, offshore endowments, domestic and foreign policies, six basic models were offered to describe the alternative approaches which might be adopted.

The second part of this chapter provided a discussion of the possible advantages and disadvantages to countries which can develop offshore oil production and to those which cannot. It is reasonably certain that the states with control over offshore hydrocarbon resources will have the opportunity to realize substantial benefits from these resources. The resultant activity is not likely to benefit substantially the states without access to offshore oil and, in certain cases, the non-offshore oil countries could even experience adverse economic effects.

The distribution discussion in Chapter 1 indicated that fifty-five countries (twenty-five no-potential and thirty low-potential countries) with incomes less than $500 per capita—including most of the United Nations' least developed countries—have very little or no promise for offshore oil production. Furthermore, there is virtually no chance that the favorable effects of production of oil from the offshore areas of other developing nations will exert an appreciable beneficial economic impact on the non-oil producing countries. These two points suggest that neither nature nor economic forces can be expected to implement the common heritage concept in the exploitation of offshore hydrocarbon resources.

If the common heritage concept, introduced by Dr. Arvid Pardo in 1967, endorsed by the United National General Assembly in 1970, and prominently mentioned in the United States draft treaty of 1970, is to be realized in any material sense by the world's poorest countries, it will be necessary to modify the implications of the natural distribution of offshore hydrocarbon resources in such a way that those nations can actually realize some benefits from mankind's common heritage.

NOTES TO CHAPTER 3

1. For a documentary work which attempts to detail the relative advantages of international oil companies over LDC national companies, see *Petroleum Exploration: Capital Requirements and Methods of Financing*, United Nations, Department of Economic and Social Affairs (New York: United Nations, 1962), cited in Tanzer, *Political Economy*, p. 391, n. 12.

2. For example, see U.S. Federal Energy Administration, International Energy Affairs, "The Relationship of Oil Companies and Foreign Governments," Washington, D.C., 15 February 1975. (Mimeographed.)

3. Michael Tanzer details the erosion of major international oil company influence in the decades of the 1950s and 1960s. Besides increased nationalism, he attributes the decline to increased competition from "international minors" and to practices by European national companies which cracked the technology monopoly of the international majors.

4. "Brazilian Oil Field," *Washington Post*, 6 December 1974, p. A9.

5. Offshore drilling tends to be fundamentally much more difficult than onshore drilling because the former requires slant and directional drilling as a substitute for changing the location of the entire drilling apparatus. Slant (the hole is not perpendicular to the surface) and directional (original direction of the hole is changed during the drilling process) drilling requires highly skilled and experienced personnel and specialized equipment.

In 1973 it was estimated that over 80 percent of the fleet of 280 mobile drilling rigs would be American-owned and -operated by 1975. The six largest drilling companies, all American, would be operating about fifteen rigs each, while their four largest European competitors would own an average of only five rigs each. The largest European firms would not exceed a size of ten rigs before 1985 (Verschure, "How Long," pp. 136–41). The European lag is attributed to the highly technical nature of offshore drilling and the need for a sufficient aggregation of capital to allow for the formation of multiple-rig firms with their inherent risk advantages. (The risk element for hired drilling rigs is not directly linked to the risk of not discovering petroleum but to the risk of rig nonutilization.)

6. See page 44 for an example of a basic cost computation. In practice, prospective investors will evaluate a potential investment with respect to the discounted income flow projected to result from the investment. To make the investment decision, they may either adjust downward the expected income stream or they may increase the profit component of the income flow in excess of a normal return on capital. For example, the cost calculations in Table 2–4

were based on a 12 percent interest rate, which essentially represents the return on an investment in a relatively low risk circumstances. As the estimate of riskiness, or margin of estimated error, increases, the rate would be increased from 12 percent. Tanzer notes that "a major new innovation (since the late 1950s) in computing the profitability of projected investments has been use of the 'discounted cash flow' (DCF) method. Formally, this method defines the rate of profit as that discount rate which, when applied to the projected future stream of earnings from an investment project, makes the discounted value of this stream equal to the initial investment" (*Political Economy*, p. 32).

7. The means of financing may be expressed in terms of the surplus of domestic income over domestic consumption or the excess of foreign exchange earnings over foreign exchange expenditures. The former expression—savings—is used although reference to a foreign exchange surplus would be a suitable alternative. There is an extensive literature in economics called "two-gap" analysis which is relevant. The literature takes the form of a number of two-gap models which are essentially concerned with the excess of investment requirements over domestic savings and/or the excess of import needs over export earnings, given a target growth rate. The objective of two-gap analysis is to quantify the need for foreign resources to balance domestic resource shortages in the course of achieving a certain growth rate. While the two-gap approach can be criticized, it could be one way of learning more about a developing economy's relative ability to finance offshore operations. For elaboration and criticisms of two-gap analysis, see H.B. Chenery and A.M. Strout, "Foreign Assistance and Economic Development," *American Economic Review* 56 (September 1966):679–733. See also K.B. Griffen, "Foreign Capital, Domestic Savings and Economic Development," *Bulletin of the Oxford University Institute of Economics and Statistics* 32 (May 1970):355–69; Ronald I. McKinnon, "Foreign Exchange Constraints in Economic Development and Efficient Aid Allocation," *The Economic Journal* 74 (J (June 1964):388–409; and G.F. Papanek, "The Effect of Aid and Other Resource Transfers on Savings and Growth in LDCs," *The Economic Journal* 82 (September 1972):934–50.

8. See Tanzer, *Political Economy*, Chapter 8, for a more complete discussion of IBRD policies regarding petroleum development in developing economies. In view of changes that have taken place since 1969, it is plausible that the policies described by Tanzer may be altered. See also Andrew Shonfield, *The Attack on World Poverty* (London: Chatto and Windus, 1960).

9. "Oil Industry in Poor Nations May Get World Bank Loans," *Washington Post*, 15 March 1976, p. D9.

10. It is immaterial whether the developing country oil enterprise conducts most of the work itself or subcontracts portions to specialists; it must still be competent in the lender's judgment to oversee the total program.

11. "Major Oil Developments Programmed to Make Peru a Net Exporter within Two years," *IMF Survey* 4 (28 April 1975):125–27.

12. U.S. Federal Energy Administration, "Relationship of Oil Companies."

13. There remains the possibility of a developing country which has the offshore potential but lacks the technology and financial resources to pursue an independent path. The country nevertheless refuses, for political or ideological

reasons, to permit the minimum amount of foreign participation necessary to allow production. While this course would not be rational in an economic sense, it is quite conceivable.

14. Industry sources.

15. See pp. 45–47 of Chapter 2 for a discussion of gestation periods.

16. These effects can be expressed explicitly in equation form. The net increase in foreign exchange (FE) takes into account gross foreign exchange earnings from petroleum sales (X) and foreign exchange savings (S) equivalent to exchange earnings from foregone oil imports less wages, dividends and interest transferred abroad (D) less imports of goods and services used as inputs (M) plus net investment (I). The net foreign exchange increase is expressed in the equation

$$FE = (X + S) - D - M + I.$$

It is further noted that a portion of the foreign exchange earnings may take the form of government tax revenues on petroleum production. There are a number of kinds of taxes but they all are essentially either payments for production and/or production-related activities. Payment may be in cash or in kind in which case the government may either sell the oil back to the producing company or market it independently. Regardless of form, the foreign exchange effect of government tax income from oil production is equivalent to exchange earnings from ordinary export sales. The second effect called the "resource availability effect," takes into account "new investment as a contribution to increased output that will increase future foreign exchange receipts." Resource availability (RA) is expressed as

$$RA = (X + S) - D - M' + I.$$

The symbol M' is imports for current production and capital replacement. The equation does not, but probably should, include investment expenditure from domestic sources. The third kind of quantifiable effect from offshore production, called the "direct domestic input effect," is expressed in equation form as

$$P_d = W + L_p$$

in which W is wages to local labor and L_p is purchases of local materials ("nonfactor inputs"). Raymond F. Mikesell *et al.*, *Foreign Investment in the Petroleum and Mineral Industries: Case Studies of Investor-Host Country Relations* (Baltimore and London: The Johns Hopkins Press for Resources for the Future, Inc., 1971), pp. 23, 24. These equations are simplified versions of a more elaborate analysis by Markos Mamalakis, "Contribution of Copper to Chilean Economic Development, 1920–67: Profile of a Foreign-Owned Export Sector," in Mikesell *et al.*, *Foreign Investment*, Chapter 16.

17. See Albert O. Hirschman, *The Strategy of Economic Development* (New Haven and London: Yale University Press, Yale Paperbound, 1958), for an extensive discussion of forward and backward linkages.

18. While there is copious literature on several aspects of international reserves, the relationship between LDC export earnings (level and fluctuations) and reserve behavior has not been extensively explored. Some insights may be gained however by reviewing William L. Hemphill, "The Effect of Foreign Exchange Receipts in Imports of Less Developed Countries." *IMF Staff Papers* 21 (November 1974): 637–77; and M. June Flanders, *Toward Assessing the Need for International Reserves*, Essays in International Finance no. 58 (Princeton: Princeton University Press, 1967).

19. For an introduction to the debate, see Constantine Glezakos, "Export Instability and Economic Growth: A Statistical Verification," *Economic Development and Cultural Change* 21, pt. 1 (July 1973): 670–78; P. Kenen and C. Voivodas, "Export Instability and Economic Growth," *Kyklos* 25, fasc. 4 (1972): 791–804; A. McBean, *Export Instability and Economic Growth* (Cambridge: Harvard University Press, 1966); U.S. Department of State, Agency for International Development, *Export Earnings Fluctuations and Economic Development: An Analysis of Compensatory Financing Schemes*, by Thomas Morrison and Lorenzo Perez, A.I.D. Discussion Paper no. 32 (Washington, D.C.: Agency for International Development, September 1975); and Odin Knudsen and Andrew Parnes, *Trade Instability and Economic Development: An Empirical Study* (Lexington, Mass.: D.C. Heath and Co. 1975).

20. See Hla Myint, "The Classical Theory of International Trade and Underdeveloped Countries," *The Economic Journal* 68 (June 1958): 317–37; Gunnar Myrdal, *Economic Theory and Under-developed Regions* (London: Gerald Duckworth & Co., Ltd., 1957); Raul Prebisch, "Commercial Policy in the Underdeveloped Countries," *American Economic Review: Papers and Proceedings of the Seventy-first Annual Meeting of the American Economic Association* 49 (May 1959): 251–73; and Hans Singer, "The Distribution of Gains between Investing and Borrowing Countries." *American Economic Review: Papers and Proceedings of the Sixty-second Annual Meeting of the American Economic Association* 40 (May 1950): 473–485.

21. The relevant concern is the rate of extraction. This rate in turn, is a direct function of the discount rate. For a complete discussion, see Harold Hotelling, "The Economics of Exhaustible Resources," *The Journal of Political Economy 39 (April 1931)*: 137–75; and Robert M. Solow, "The Economics of Resources and the Resources of Economics," *American Economic Review: Papers and Proceedings of the Eighty-sixth Meeting of the American Economic Association* 64 (May 1974): 1–14.

22. In a standard economic approach, the rate of extraction is related to the rate of interest and the rate of capital gains (the rate of increase in the price of the natural resource). As the rate of interest declines with respect to the expected capital gains rate, application of economic criteria would call for a decrease in the rate of extraction. Absolute prohibition of mineral production is unsound so long as the present value of the resources produced is greater than the discounted future value of production. Even then, total cessation of production may not be economically rational. For a more detailed discussion see Hotelling, "The Economics of Exhaustible Resources;" Solow, "The Economics of Resources"; and Mikesell, *Foreign Investment*, p. 18.

23. Ronald I. McKinnon, "International Transfers and Non-Traded Commodities: The Adjustment Problem," in *The International Monetary System and Developing Nations*, ed. Danny M. Leipziger (Washington, D.C.: Agency for International Development, forthcoming).

24. The key conditions for such an ignominious result are that the size of the international transfer be modest relative to the pay increase going to a small part of the labor force and inducing a wage increase for "a much larger number of workers in non-tradable and pseudo non-tradable activities, increasing the capital/labor ratio and expelling workers back to the unprotected agricultural (tradable goods) sector" (*Ibid.*).

25. John W. Devanney, III, "Key Issues in Offshore Oil," *Technology Review* 76 (January 1974):21–25. On oil pollution issues, see Kash, *et. al.*, *op. cit.*, Appendix A, B and Wendy Garling and Danny M. Leipziger, "Pollution in the Marine Environment: Interests of Developing Countries," unpubl. paper, Agency for International Development, U.S. Dept. of State (April, 1975).

Chapter 4

The Disposition of Benefits from Offshore Hydrocarbon Production

The preceding chapters have presented the locational distribution of offshore hydrocarbons and the technical and economic aspects of their exploitation. In this final chapter on undersea petroleum resources, the allocation of revenues generated from offshore petroleum activities is examined. For the greater part, revenue generated from the offshore petroleum installations in place in 1975 are not considered to be within the scope of the discussion. The examination concentrates on areas seaward of 12 nautical miles from shore.

FROM INTERNATIONALIZATION TO NATIONALIZATION

In Law of the Sea discussions throughout the 1960s, the dominant theme was that benefits from exploitation of mineral resources in and on the seabed in international areas should be shared by the international community. Yet the main thrust of the present Law of the Sea debate pertaining to offshore hydrocarbons is not how exploitation might best be implemented to achieve an international distribution of benefits, or even whether states should be able to claim large tracts of oil-bearing seabed. It is now a basic working premise that certain states—those which happen to border on the world's oceans—will acquire exclusive exploitation rights to virtually all of the hydrocarbons adjacent to their coastlines. To illustrate how far from internationalization the discussion has moved, the most intensive debates are now expected to center on the bilateral negotiations between pairs or sets of coastal states to settle specific marine boundary disputes and to establish mutual recognition of unilateral claims.

Perhaps the high point in the trend toward some form of international sharing of seabed resources was reached with the release by the United States of the "Draft United Nations Convention on the International Seabed Area: Working Paper" in August 1970. This working paper suggested that the band of seabed situated seaward of 12 nautical miles from the coast and extending to the base of the continental slope be designated a trusteeship zone and that one-half to two-thirds of the revenues generated in the zone be turned over to an international authority. The spirit of the working paper was consistent with the general spirit of the numerous pronouncements made by private groups concerned with the seabed, with Dr. Arvid Pardo's "Common Heritage of Mankind" address to the United Nations in 1967, and with President Johnson's declaration in 1966 that "We must insure that the deep seas and the ocean bottom are, and remain, the legacy of all human beings" [1]. It is not clear when or how the working premise of internationalization of the seabed was reversed and became one of ratifying nationalization. It is clear that between 1970 and the close of the Third Law of the Sea Conference in August of 1974 the change had taken place, the degree of change was extreme, and the change was irreversible.

The reason why the reversal took place and the reason for the extent of the change are not readily identifiable. Probably there were a number of factors which contributed to the new direction at Caracas. From the outset several coastal states, especially in Latin America, favored extension of their territorial seas to 200 miles [2]. By the early 1970s there had been a substantial accumulation of evidence on the existence of geological formations with high petroleum potentials beneath the ocean floors. At the same time, there was an increasing technical capacity to find and exploit these formations. Perhaps the greatest weight was thrown onto the scales when the price of oil quadrupled in a few months' time and when petroleum supplies to many countries were curtailed in the fall of 1973. It seems unlikely that such traumatic events could have failed to exert a strong influence. By the summer of 1974 all but a handful of countries were confronted by greatly enlarged oil import bills. All nations, regardless of import status, were faced with greatly increased petroleum costs and the knowledge that the oil trapped in the continental margin off their shores constituted a nearby, uninterruptable supply of a fundamental, strategic material whose commercial value had just increased fourfold. It is difficult to imagine that the government officials of coastal states who were preparing for the Caracas conference in the spring of 1974 could seriously contemplate any negotiating position which did not lean toward the acquisition of maximum rights to offshore petroleum. Whatever the reason or combination of reasons, the reversal which materialized in Caracas was so complete and so frictionless that it could hardly be called a revolution. Yet it was a revolution in which nationalist sentiments overwhelmingly dictated the new course of the Law of the Sea negotiations as they pertained to the rights of access to seabed hydrocarbons.

The relatively minor attention paid to the subject of revenue sharing provides another indication of the completeness of the move toward nationalization of the seabed. The concept of revenue sharing from seabed hydrocarbon exploitation was not a new idea. It had been a major component in the proposal put forward in the 1970 United States Working Paper. There were two different revenue-sharing schemes introduced at the meeting in Caracas, but neither was developed as a *quid pro quo* to the move to extend national rights to encompass approximately 30 percent of the ocean floor [3]. The fact that a part of the international community was suddenly to be excluded from an important segment of seabed resources was an historic development; the fact that compensation did not evolve as an important issue is at the very least noteworthy.

The absence of substantial debate over revenue sharing from seabed hydrocarbon exploitation is puzzling but not incomprehensible. Coastal states, developed and less developed, outnumbered non-coastal states by a two-to-one ratio. Perhaps more significant, a large number of the non-coastal states tended to be at a disadvantage relative to many of the coastal states, which possessed large negotiating staffs backed by extensive legal and technical resources. The end result was that the coastal states, while not acting in collusion, nevertheless reacted in unison to the common stimuli—higher oil prices and embargo, or the implied threat of future embargoes—and effectively presented a similar, firm position in favor of increasing their own control over the continental margins.

It is not likely that the trend in favor of nationalization was a total surprise to the non-coastal states; however, the large number of states supporting the trend and the relatively short time span (ten weeks) afforded at the Caracas conference did not provide enough time for the non-coastal states to develop an effective counterproposal in terms of compensation. Finally, the prevailing attitude at Caracas was that it constituted a preliminary session at which states would concentrate on establishing their basic positions rather than engaging in substantive negotiation on technical subjects. The second session of the Third United Nations Law of the Sea Convention, held in Geneva in March and April 1975, did not witness any substantially new developments, although one revenue sharing proposal was set out in more detail and the geographically disadvantaged countries were slightly more voluble.

The question of coastal state economic jurisdiction was essentially answered at Caracas. First, coastal states are likely to acquire at least the right to hydrocarbon resources extending seaward to 200 nautical miles from their shores. Second, there is a reasonable chance that the entire continental margin including those portions beyond 200 nautical miles will be divided among the coastal states [4]. The difference between the two alternatives is relatively small in any event—a difference of approximately 284 billion barrels of production potential, or 12 percent of the total amount of hydrocarbons estimated to be contained within the margins.

For the non-coastal states, and especially the poor non-coastal countries, the outcome of any debate over economic jurisdiction between 200 miles and the edge of the margin is particularly insignificant because these poor states are not likely to exploit promising formations beyond 200 miles before the firms from the more wealthy coastal states establish claims. Furthermore, if gains from exploitation of the hydrocarbon resources beyond 200 nautical miles were to go to the international community, the amount of potential revenue is likely to be small, not only relative to the gains realized by the coastal states but in absolute terms as well because the expense of exploitation beyond 200 nautical miles is much greater, thereby discouraging exploitation. In the exceptional cases where exploitation at the extreme edge of the continental margin might be feasible, the net return could still be expected to be very small.

THE RATIONALE FOR REVENUE SHARING

In the following paragraphs a new rationale for sharing the benefits from continental margin resources (revenue sharing) with geographically disadvantaged states is introduced. The first and primary issue is determination of whether grounds for compensation exist. The focus is whether the benefit to excluded states is proportionately as great as it is for coastal states. Two points can be made. First, in a strategic sense, there is potential for benefit to the excluded states if new offshore production results from an addition to the number of producing states. Such an addition increases the number of petroleum suppliers, thereby decreasing the chance of total embargo and possibly decreasing the effectiveness of an embargo, should one occur. Second, on the economic side, excluded states still benefit because of the greater number of producers and because the relative increase in output may cause downward pressure on petroleum prices.

But there is a very substantial economic benefit which excluded states cannot acquire: they are not allowed to collect any of the economic rent or economic profit which is usually a prominent feature of petroleum production [5]. The term "economic rent" in this context is taken to mean the difference between the selling price of the crude product and total production costs, including a reasonable return on investment [6]. For example, if the average total cost per barrel at the wellhead were $4 and the price per barrel at the wellhead were $10, then the rent element would be $6.

Because the rent element in crude oil production can be substantial, economic rents tend to be associated with petroleum production as one of its prominent characteristics [7]. For example, the total cost of producing crude oil in the Middle East is somewhat less than $1 per barrel, whereas the price of the crude at Persian Gulf loading ports ranges around $10—a rent of $9 per barrel. Similarly, if coastal states were to gain exclusive rights to large amounts of offshore petroleum resources, they would also obtain the exclusive right to extract the rent element from these resources [8]. While the Middle East example is cited, it must be kept in mind that because of the higher costs of off-shore production, the amount of economic rent can be expected to be substantially less, and in some cases the total cost of production per barrel may just equal the price per barrel.

There are a number of arguments which suggest that all nations have a claim to the rent element of deep-water offshore oil. Two of these arguments can quickly be dismissed. The first is that a claim is justified because over the last ten or twelve years statements have been issued by various United Nations groups and by governments to the effect that deep-water offshore hydrocarbons are a common good. The pronouncements and proposals in the 1960s, however, were produced as preliminary or exploratory work and, regardless of the findings, do not constitute an obligation. As a second proposition it has been argued that acquisition of certain exclusive economic rights in a region initially under international control calls for compensation to the excluded countries by the nations which gain the rights. Opponents of this view would point out that there is no historical precedent for this. For example, non-coastal states were not compensated when the 3-mile territorial sea limit was introduced several centuries ago, nor was compensation an issue when national jurisdiction was extended beyond the 3-mile limit and claims were made on resources in and on the continental shelf following World War II.

One of the most interesting propositions which would appear to support a position in opposition to a claim by non-coastal states has been advanced by Dr. Hollis D. Hedberg [9]. While not explicitly addressing the compensation issue, Dr. Hedberg's argument provides a rationale whereby coastal states may establish a direct claim on the sections of coastal margins adjacent to their shorelines and then, based on this direct claim, preclude any claims by non-coastal states.

Dr. Hedberg suggests that there is a definitive and logical solution to the dispute among the proponents of various proposals for seaward boundaries involving distance from shore, water depth, or some combined criterion. He forcefully argues that all distance and depth criteria are arbitrary and that references to one frequently cited criterion, the 200-meter isobath, "results from the classic but erroneous assumption that the 200-meter isobath corresponds with the water depth at the edge of the shelf." Furthermore, Hedberg argues, boundaries based on distance or depth would not likely be acceptable to coastal states. A distance or depth boundary which did not include the entire continental margin adjacent to the coastal state would "deprive some coastal nations adjacent to broad continental shelves of much of what they have been accustomed to regard as their own natural territory (e.g., Argentina and the USSR)." A water-depth criterion would allow only a very thin strip of sea for steep-slope coastal states such as Chile or Peru. A boundary based on depth or distance is impractical because the reference points are sometimes ambiguous (such as the shoreline of a low-lying coastal area), and they are subject to constant change from natural causes as well as subject to "adjustment" by coastal nations [10].

There is a boundary criterion, Dr. Hedberg suggests, which is not arbitrary and which would be satisfactory to coastal states and "in principle be far more palatable to land-locked or shelf-locked countries": the base of the continental slope, i.e., the edge of the continental margin, which would be used as a *guide* for specification of the boundary. The base of the slope would be used to establish a "zone" through which the actual boundary line, consisting of "points fixed by coordinates of latitude and longitude and connected by straight lines or arcs," would be designated by the coastal states [11].

The reason for the selection of the base of the continental slope as the definitive feature is very persuasive:

> The surface of the earth consists on the one hand of positive elements—continents, islands, banks, and other topographic eminences; and on the

> other hand of negative elements—the great central ocean basins and their depths. . . . The submerged continental shelf and the submerged continental slope surrounding a continental land mass belong to that continent and not to the ocean basins between the continents. Likewise, the insular shelf and insular slope surrounding an island area are part of the island, not of the ocean basin [12].

In addition to the distinct geomorphic difference between the continents with their submerged shoulders and the abyssal plain of the ocean floor, there is a geological difference as well. "The base of the continental slope . . . also approximately marks a fundamental lateral change in the character in the rocks of the Earth's crust—the change from the relatively low density sialic rocks of the continental crust to the relatively higher density mafic rocks of the oceanic crust. . . ." Thus, Dr. Hedberg concludes, "the base of the continental slope is a natural boundary geographically, geologically and geomorphically" [13].

The argument by Dr. Hedberg is based on physical facts and as such constitutes a much stronger basis for coastal states to claim the continental margins and their resources than the rationale that technical advances have made resources within the margins accessible and therefore some states ought to be able to exploit them. The exposition of the geographical, geological and geomorphic continuity (hereafter referred to as the "geo-link") between coastal state and continental margin is not complete, however. Given the fact that "the submerged continental shelf and the submerged slope surrounding a continental landmass *belong to that continent*" (italics added), it is not established that the coastal states have an *exclusive* claim on their adjacent coastal margins unless it can be established that there is no geo-link between coastal states and adjacent non-coastal states. Of course, in no sense are coastal states and non-coastal states on a continent separated in the Hedberg "geo" sense by national boundaries [14]. Thus the geo-link which gives coastal states a claim to the international areas of the continental margins gives all states a claim, including non-coastal states. The Hedberg proposition therefore constitutes a very strong argument for a claim by all of the nations of a continent to the underwater portions of that continent. But it in no way establishes a unique claim for the coastal states alone.

The problem with the Hedberg geo-link thesis is not in its logic, but in its implications for application. How would the claims of landlocked and other geographically disadvantaged states be defined with respect to size and location of claim? As the concept is originally put forward by Dr. Hedberg, this problem is avoided by merely extending the coastal state economic jurisdiction seaward and ignoring the fact that the states on the continent which do not border on the ocean are also part of the same geological-geomorphic unit. If, as a result of the geo-link, the claims of all the states on a continent are to be recognized—instead of just the claims of the coastal states—it becomes necessary to allocate parcels of the continental margins to all countries sharing the continent [15]. Whereas a number of schemes for allocation of claims are conceivable, the basic approach is objectionable on practical grounds because it is likely to involve an extremely complex, lengthy, and undoubtedly conflict-ridden process.

The alternative to subdivision of the continental margins among the countries of a continent is financial compensation by coastal states to the geographically disadvantaged states in return for exclusive control. This method would avoid the fundamental political and strategic complications of a physical allocation approach, and it would recognize the preeminent interests of coastal states in the areas off their coasts.

Before proceeding to a discussion of the aspects of financial compensation, a second problem emanating from application of the (expanded) Hedberg geo-link is noted. The large differences in the numbers of geographically disadvantaged countries on the continents would require that there be extensive division—financial or physical—of some continents' margins (e.g., Africa), while others with no geographically disadvantaged states would not require division (e.g., North America). In addition, it would be necessary to determine the continental identification of land-links between continents for purposes of margin division. For example, if the countries of Central America were defined as part of South America, some margin sharing with Paraguay and Bolivia would be obligatory. However, if the countries of Central America were considered to be part of North America, there would be no obligation.

If the logic of the Hedberg proposition is consistently applied, however, the conventional differentiation among continents disappears. It does so because all of the continents are one unit in a geological and geomorphic sense. None of the major land masses exists as an island completely surrounded by an abyssal plain; they are all parts of the same element. The southern tip of Africa is part of the same "positive element" of the earth's crust as is the southernmost tip of Argentina or Australia because all of the continents are linked by submerged continental margins which are themselves nothing more than submerged portions of the continents. All of the earth's continents are linked either by dry land bridges or by submerged portions (none of which lies under more than 200 meters of water) [16]. Thus, the globe's continents are nothing more than subcomponents of a single geological-geomorphic mass.

This point gives rise to a fundamental implication for the allocation of claims among countries. It essentially means that the claims of geographically disadvantaged states need not necessarily be limited to the margin of the continent on which they are located. If geological-geomorphic continuity establishes a claim by coastal states to continental margin resources in the international area, then the same rationale establishes a claim by all states to all margin resources in the international area.

In summatiion, while the geo-link argument as initially proposed by Dr. Hedberg was undoubtedly intended to establish a physical or natural basis for a claim by coastal states on continental margin hydrocarbon resources, simple extension of the argument (and it would be argued without twists or contortions) provides the basis for all states to establish a claim on offshore hydrocarbons [17]. The extension of the Hedberg thesis, however, establishes a physical basis for all countries to claim a share of offshore resources but unfortunately it does not resolve the problem; that is, universal agreement is not likely to occur simply because landlocked countries, nations with coastlines, and oil deposits happen to share the same geological unit. The extended Hedberg thesis nevertheless serves a useful purpose as a backdrop for highlighting the subjective nature of other rationales, i.e., coastal states ought to be able to appropriate the total value of offshore hydrocarbons, because they have (or can acquire) the means to extract the resources, as opposed to geographically disadvantaged states which ought to receive a share of offshore hydrocarbons in the interests of fairness and equity. The extended Hedberg thesis thus serves as a useful benchmark by which to judge the degree of impartiality in the final resolution of this twentieth century conundrum. Because an objective rationale does exist for division of benefits from hydrocarbons located in the international regions of the globe, the remainder of this chapter will focus on the economics of implications of such a division and very briefly outline selective aspects of its implementation.

The division process is referred to as revenue sharing or benefit sharing. In the context of the following discussion, revenue sharing means that coastal states, having a preeminent interest in their adjacent coastal waters, would exercise the right to determine when, where and who might extract hydrocarbon resources; but that a portion of the economic rent element in the resources, once they are extracted, would be allocated for distribution to geographically disadvantaged states.

REVENUE SHARING AND ECONOMIC DISTORTION

Revenue sharing may be viewed essentially as a tax on offshore exploitation and, as such, warrants a brief digression. In a typical taxing situation, the amount of revenue to be raised can be determined in theory and approximated in practice by relating the tax either to the benefits derived from expenditure of the tax or to the taxpayer's ability to pay [18]. The fact that the purpose of the revenue sharing scheme is to transfer revenue and not to benefit those paying the tax (the issue of who actually pays the tax is discussed below) implies that a tax-benefit comparison is not feasible. The ability-to-pay principle may be more applicable in that the tax is related to output (or profits which reflect output), which is a reference to the ability to pay. But the determination of a tax is not made totally without guidelines; ideally it should not discourage investment or alter production decisions and it should be related to the activity being taxed.

The non-distortion caveat is in a sense a straightforward prescription, but in a real world context, where all kinds of economic distortions already exist, it is not necessarily a binding constraint [19]. Onshore petroleum production is taxed; offshore operations can be expected to be subject to existing national taxes as they come into existence; and other sections of the world economy contain a large and varied assortment of economically distorting elements.

Consider two examples of how distortions might relate to an offshore revenue-sharing tax. In the first case a revenue-sharing tax might actually improve the overall allocation of resources. If it is assumed that costs are greater for offshore operations relative to onshore operations but that the taxes for onshore petroleum production are greater than those for offshore production, then proportionately more resources will be directed to more costly (in the sense of requiring a relatively greater amount of inputs) offshore exploitation. Because it requires more of society's resources to produce offshore than would be required to produce the same marginal output onshore, the tax differential leads to a wasteful allocation of resources. An equalizing tax on offshore operations could rectify production misallocation by equalizing taxes in the two areas. Whether or not the additional tax on offshore activity would be detrimental to investment remains unanswered. However, the response would not pivot solely on the introduction of the offshore tax, but would depend also on the existing aggregate level of taxation levied on offshore petroleum activity and the rate of return on investment which might be attained in other sectors of the economy.

In the second case, a revenue sharing tax could also have the beneficial side effect of slowing down the pace of petroleum extraction to a rate which is more consistent with some long-run, socially optimum production rate. The economic theory of exhaustible natural resource exploitation postulates that the rate of extraction is a direct function of the market interest rate, and because individuals tend to discount the future at a rate greater than the social discount rate (since individuals must also take into account the risk of transfers within society), the private rate of exploitation tends to exceed the socially optimum rate [20]. The problem is that without specific knowledge of the distortionary effects which exist in all other economic activities, the economic impact of a new revenue sharing tax on production and investment cannot be precisely known.

THE BASIC FEATURES OF A REVENUE SHARING SYSTEM

If any revenue sharing system were to be implemented, it would contain three interrelated core elements aside from the actual institutional-administrative apparatus: (1) the tax base, comprised of the area of continental margin over which revenue sharing is applicable and the exact form of the product being taxed; (2) the valuation base, which essentially determines the form of taxation, e.g., a tax on the value of output (a royalty), a tax in kind on the volume of output, a one-time tax on the activity, or a tax on profits; and (3) the rate of revenue sharing or taxation. The fact that revenue sharing has a specific rationale in the geo-link thesis provides some basis for guidance with respect to the area of application. Whereas the geo-link postulate does not provide a specific boundary where revenue sharing would start, presumably it would apply to the international area, that is, the ocean floor seaward of coastal states' territorial claims. It can be seen in Table 4–1, however, that the breadth of the international area beyond the territorial seas would vary a great deal according to individual coastal states' territorial sea claims.

If designation of the revenue sharing area is not to discriminate among coastal states, the landward extent of the international area for revenue sharing purposes would have to be uniform, e.g., a common distance or isobath for all countries. Uniformity of the boundary is not necessarily sufficient to insure intercountry equity, however. Consider two countries such as Chile and Argentina, the first with a margin which descends very rapidly to the abyssal plain and the latter with a margin which descends very gradually. A distance boundary of, say, 50 or 60 nautical miles would effectively

Table 4–1. Territorial Sea Claims

Breadth (nautical miles)	*Number of Countries*
3	30
4	4
5	13
10	1
12	57
18	1
30	4
50	3
100	1
130	1
150	9
200	3
Modified Archipelago	1
	128

Source: U.S. Department of State, Bureau of Intelligence and Research, Office of the Geographer, *Limits in the Seas No. 36: National Claims to Maritime Jurisdictions*, 3d rev., Washington, D.C., 23 December 1975, p. 8.

Note: As of December 1975 there are 29 independent landlocked countries (plus Byelorussian SSR).

exempt Chile from contributing to revenue sharing (assuming for the moment that Chile and Argentina would be subject to revenue sharing), whereas sharing of revenue would apply to a large portion of Argentina's adjacent margin. The situation would tend to be reversed, however, if, for example, a 200-meter-isobath boundary were adopted, although not necessarily to the same degree. To avoid this kind of equity problem, it would be preferable to set the revenue sharing boundary relatively close to shore to minimize the effect of the degree of the downward slope of the margin with perhaps the additional provision of a depth criterion to make some allowance for particularly steep-margin states such as Chile. Thus the revenue sharing boundary could be set, for example, at 12 nautical miles or the 200-meter isobath, depending on whichever line is closer to the shore [21]. Specification of such a broad revenue sharing area is also consistent with the sense of the extended Hedberg argument, which would otherwise be meaningless if any coastal state could unilaterally define the seaward limits of its exempted area. The most significant economic argument for a broad revenue sharing area is that it minimizes the possibility for distortion of investment and production with respect to location in the offshore area. Uniform application of the tax to virtually the entire international continental margin would ensure that no producer would be motivated to choose the location of operations in a way which could cause petroleum to be extracted at a site which required more inputs (but higher returns to the producer) than from a site which, for the same output characteristics, required fewer inputs (but lower returns to the producer). Thus, ideally, the revenue sharing area would encompass that portion of the continental margin (the area seaward of the 12-mile limit) which prior to 1958 was considered by most states to be an international area.

Having examined the first part of the tax base, the task remains to identify for taxation purposes the precise form of the product being taxed. This is relatively easy because the crude product at the production site—that is, the "pure" resource without any other value-added elements such as transportation or refinement—is identifiable and it has a market-determined price. Therefore, the value is ascertainable, and a tax on the value of the crude output would not inadvertently tax any other element. Value assessment of the crude product is feasible because reference prices already exist from ongoing production [22]. Thus the first element in a revenue sharing system is the tax base, comprised of the relevant area—which, it is argued on economic grounds, should be as broad as possible—and the specific product, the crude hydrocarbon at the wellhead.

The second core element in a revenue sharing system, the valuation base, essentially determines the basic nature of the tax. Although the revenue sharing rate is noted as a separate third element, specification of the value base carries certain implications about the revenue sharing rate, i.e., a profits tax, other things being equal, would be expected to be greater than a gross royalty. Thus, in the following discussion, whereas tax rate levels cannot be explicitly specified, each type of tax carries implications for the rate level. There can be almost any number of variations of any single type of tax or it is possible to combine two or more types of taxes in their various forms. In the following discussion, three basic types of taxes are sketched with the intention of noting their more salient features with respect to offshore hydrocarbon production, and some specific variations are outlined.

The first type is a tax on revenue remaining after deduction of all costs, including a return to capital (a return which could be adjusted for risk). The term "profit tax" will be used to identify this tax form, but the term is not meant to apply to normal returns to investment. The profits tax is one of the least distortionary methods of taxation, but it is also one of the most difficult taxes to formulate and administer [23]. In its purest form, the tax rate would be 100 percent of economic rent, with a portion of the rent tax payment being returned to the producer proportional to the producer's share of the global geo-unit. Thus the effective rate could be expected to vary for each country producing offshore hydrocarbons. A tax on profits has the advantage of not adding to unit costs of production. Therefore, the tax would allow production to increase to the economically optimal level where the additional cost of the last unit produced just equals the additional revenue earned by that unit [24].

The second point in favor of a profit tax is that it does not increase the initial cost of operations and, if the allowance made for the return on investment is comparable to the returns on alternative investment opportunities, then investment in offshore operations would not be discouraged. Finally, a profits tax is flexible with respect to cost variations, changing costs over time, and changing output prices. In addition, a tax on profits is directly related to economic rent, the sharing of which is the point of the whole exercise. Unfortunately, a profit tax also has significant disadvantages. The most prominent drawback is definition of cost items, cost concepts and other accounting elements in a manner which allows a satisfactory separation of factor costs and economic rent. Even if this substantial objective can be achieved, there remains the difficulty of either trusting individual national accounts for cost and income statements or establishing some international machinery to audit the books of offshore operations. There are a plethora of ways to manipulate costs to minimize profits, and for some countries the ideological implications of a profit tax would probably not be acceptable. A profit tax, which seems well suited on economic grounds, would probably be among the most difficult taxes to administer in practice.

A variation on the profit-tax approach is taken from a proposal by R. Garnaut and A. Clunies Ross, who address the problem of taxing natural resource extraction under conditions of uncertainty and risk aversion [25]. The authors note that "the complexity of the 'rent' concept and the government's ignorance in any particular case of the supply price of investment make the balancing of revenue loss through deterrence of new projects extremely difficult" [26]. Their objective is to devise a taxation system which does not increase costs (and therefore the risk of failure) or make returns unacceptably low while still allowing the resource controller "a generous share . . . if returns turn out to be high" [27]. Garnaut and Ross propose a "Resource Rent Tax," which would work by varying the tax rate with the rate of return. Collection of the tax would commence when a specified threshold internal rate of return on total cash flow had been attained. Then additions could be added to the tax rate as specified threshold internal rates of return are realized over time. The

Resource Rent Tax is essentially a system of progressive taxation similar in principle to the approach used in calculating United States personal income taxes. While the Garnaut and Ross Resource Rent Tax is perhaps superior to other taxation methods in terms of maximizing tax revenue flow and minimizing interference in investment and production decisions, it retains the major disadvantages listed for the constant profits tax except, because the Resource Rent Tax is somewhat more sophisticated, the degree of problems in practice could be expected to be even greater.

A second method of taxation involves a fixed payment, lump-sum or periodic. There are several variations, but only four are noted: the producer pays a given amount set by the collector, the collector accepts an amount determined in a bidding process or the payments, either set by the collector or in a bidding process, are made periodically. This general method has the advantages of being relatively simple to administer, requiring minimum involvement by the collector in producers' affairs and, because it is not related to the rate of output, it would allow each producer to pursue his optimal output path. From the viewpoint of aggregate output, this type of tax would increase cost and thus the minimum acceptable expected rate of return, thereby tending to deter investment. The periodic payment would be less of a deterrent to investment because the investor can "cut his losses," in the case of a poor resource showing, by relinquishing the area.

From the collectors' point of view, the fixed payment is an inferior means to capture a proportionate share of economic rents because insufficient information and uncertainty about future conditions make payment level determination an extremely imprecise exercise if the collector is setting payment levels. Similarly, potential producers under a bidding regime tend not to be willing to risk making the substantial investment in the detailed exploration which is necessary to estimate reserves with narrow confidence limits prior to obtaining assurance of exploitation rights. Thus, given the limited knowledge about resources at the time bidding takes place, there is a strong tendency to discount heavily advance payments. If the payment were associated with something other than a permit to exploit an area, for example, with some production features such as platforms or wells, the revenue side would still encounter the same risk-discounting problem and there would be a tendency for a less-than-optimal configuration of production apparatus.

Richard Cooper, recognizing that offshore petroleum rent taxation schemes often fail to accomplish adequately their objectives, has suggested a two-step approach which "offers a reasonable approach to the extraction of scarcity rents that may not exist." The first step would consist of a process whereby companies would bid for the right to extract hydrocarbons from an offshore site (after a period of unencumbered exploration); then a profits tax would also be levied to share in "any exceptional earnings that were not anticipated at the time of the bidding" [28].

Several kinds of taxes may be grouped under the general term "output tax," the third form of revenue sharing. A tax on output could take the form of a production sharing system, in which a portion of output is earmarked for sharing with non-producers and the output, or proceeds from the sale of the output, could be distributed. As a second variation, the earmarked output could be repurchased by the producer under a "buy-back" arrangement. The first variation is relatively cumbersome, and if the petroleum were actually to be delivered to receiving countries the system could involve appreciable overhead cost. Finally, a more literal application of the production tax would take the form of a payment or royalty corresponding either to the volume of output (a specific tax) or to the value of the output (an ad valorem tax). A specific tax would be relatively easy to administer, but it would not change in response to price or cost changes. An ad valorem tax would be more responsive, but it could involve some administrative difficulties such as in the case of vertically integrated firms (e.g., the producer of crude is the same firm which refines the crude), which might use artificially low transfer prices. This particular problem could probably be overcome by using some exterior reference price. An ad valorem tax combines the advantages of being directly related to the stream of benefits with the potential for being a relatively simple, low-overhead revenue sharing device.

All output taxes share the major disadvantage of increasing unit production cost and thereby reducing the producer's optimal output level. This has a double-edged adverse effect: it withholds petroleum production that otherwise would be economically viable and, because production is attenuated, so too is the amount of tax revenue. The greater the production cost differences, the greater the effect of an output tax on production, and offshore petroleum production is notable for exhibiting large cost differences among production sites. In the latter part of this chapter, a sales tax is applied to a hypothetical offshore area to illustrate the output and revenue effects.

Because an output tax discourages production and a tax beyond a certain level causes total tax revenue to decline, determination of the "correct" tax level (either in terms of minimizing output losses or maximizing revenue generation) is crucial, but unless a substantial amount of cost information is available, determination of the correct rate is unlikely. Thus the output tax, which first appears to be relatively simple to construct, takes on an appreciable degree of complexity.

Revenue sharing could also be accomplished by means of a joint operation in which a consortium of non-producing countries joins a national or international company to exploit offshore. The return to the non-producer consortium would then include an implicit rent element (assuming no other sharing taxes exist to decrease the total income flow). This approach as stated, however, effectively assumes away the problem. It assumes that all non-producing countries have or can acquire the financial and technical resources to enable them to become part of a joint operation. Also, in order to provide a fully equitable solution, each member of the consortium would have to invest in proportion to its claim (by whatever formula that is determined) on continental margin resources.

Whereas many non-producing nations lack the capability to become co-investors, it is conceivable that offshore producers could make payments to the non-producers in the form of equity shares in offshore investments. This option would have the effect of automatically making an investment decision for lower income countries, and it distributes risk. Because the sharing of rents would be in the form of stock, dividends from these stocks would not include any economic rent because it is already being paid out in the form of stocks. This approach would come very close to bringing about global participation in exploitation of offshore petroleum resources in the international areas, but it still faces the familiar problems of determination of the rent element and its allocation among nations.

One writer has sought to combine revenue sharing with revenue distribution by varying the revenue sharing rate with respect to per capita gross domestic product (GDP), distance of the production site from the coast and water depth [29]. The revenue rate is designed to increase as a direct function of distance, depth and per capita GDP. This approach provides a response to the issue of what to do about poor producing countries: they share, but at a reduced rate. It has a substantial disadvantage in that it would introduce an appreciable distortion in the choice of production site location. Specifically it would tend to cause producers to discriminate heavily against production in deep water areas distant from shore.

As a last variation on taxation schemes, a two-level system could be initiated using an ad valorem tax to determine the international obligation, but the tax would be collected by the national government by means of a corporate income tax and the national government would in turn make the payment to the international collector. This would circumvent the awkward situation of an international entity taxing a national firm and, because the producer is paying taxes on income and not on output, his production decision remains unaffected. While attractive, the two-level approach is not without disadvantages. First, there is no reason to believe that, for the same period, the price of oil times the quantity of oil produced times the tax rate will equal the profit tax revenue unless profit taxes are adjusted ex post. With an ex post adjustment, however, the system begins to take on the flavor of an ad valorem tax from the producer's point of view. In the event that the international obligation exceeds the income tax payment, assuming no ex post adjustment, the national treasury sustains a loss.

There is an additional and perhaps much more significant problem, however. If the payment by the national treasury to the international authority comes out of ordinary domestic tax revenues, it would become extremely difficult in an appropriation process to differentiate the revenue sharing payment from the government's budgeted contribution to international economic development. Thus ordinary international contributions could be reduced by an amount equal to the tax payment. In this event, shared revenues would not be additional to the flow of economic development funds, and the effective net change in the flow of funds to low-income countries would be zero. Revenue sharing would exist only on paper. The additionality problem is not uniquely related to the two-level approach. However, additionality becomes particularly crucial whenever revenue sharing funds are collected as ordinary domestic tax revenue, the expenditure of which usually requires explicit approval in an appropriations process.

The most recent revenue sharing scheme was introduced at the 1975 Law of the Sea negotiations in Geneva. It was proposed that revenue sharing be applied only to hydrocarbon production from sites located beyond 200 miles from land. The revenue sharing rate would be 1 percent on the value of the crude product at the wellhead commencing in the sixth year of production and increasing by 1 percentage point for each of the following four years [30]. This revenue sharing scheme is indicative of the trend in the Law of the Sea negotiations away from internationalization of seabed resources toward their nationalization. It will be recalled that Lewis Weeks estimated that about 284 billion barrels of petroleum, or approximately 12.5 percent of total offshore recoverable resources, can be expected to be found in the area seaward of 200 miles from shore. Weeks adds that, for the total offshore area, less than 25 percent will be found in highly productive areas. Because of the high cost of extraction in the more distant and deeper waters, the productivity of the area gains increasing prominence, thus suggesting that only a fraction of the 12.5 percent may be commercially recoverable.

With the exception of one promising area off the coast of Labrador, virtually all of the seabed beyond 200 miles lies under more than 200 meters of water. One can only speculate on what the potential revenue yield might be, but it is difficult to be optimistic. Exploration and development would presumably occur in the areas where long-run unit output costs tend to be lowest, which means that development of the area containing 90 percent of potential offshore production, which is closer to shore and at shallower depths, would tend to precede exploration and production in the extreme outer fringe of the continental margin. If a sufficiently large field were to be discovered in the fringe area, six more years would pass before a dollar of revenue would be generated, and revenue sharing would reach its peak rate of 5 percent ten years after the beginning of production. The productive life of an oil field before secondary recovery methods have to be employed is about ten years, and production tends to start declining after the first five years of production. Thus the stream of revenue to geographically disadvantaged states (or potential revenue-receiving states, however defined) would start approximately midway through exploitation of the last offshore area to be developed and reach a peak of 5 percent at about the time the field is exhausted. In other words, the revenue-sharing base would already be in decline when the 1 percent rate is introduced, and the base might disappear altogether after approximately five years.

Assuming a random distribution of field size in the continental margins with respect to distance from shore, the increasing costs of exploration and production relative to water depth and distance would imply that proportionately less of the hydrocarbons beyond 200 miles would be developed than are developed in areas less than 200 miles from shore. It is conceivable that further development of new undersea technology such as subsea completion systems may eventually change the cost characteristics of underwater exploitation. But as the 1975 Geneva revenue sharing proposal stands, revenue sharing will not occur to any extent for perhaps the next two decades, and then the revenue-sharing base will have started to decline. Finally, it is not unrealistic to speculate that, in view of the extended time span and the fact that the hydrocarbons beyond 200 miles generally constitute what must be the world's most costly hydrocarbon resources to exploit, the development of petroleum substitutes may preclude exploitation altogether and thereby eliminate even a trace of revenue sharing.

The preceding section has touched upon the effects basic kinds of taxes might have on offshore hydrocarbon production and reviewed a few alternative approaches to taxation. Clearly, no position is adopted which favors a specific approach to taxation. It would be hoped that an international conference convened to set out a revenue sharing system would also hesitate to go beyond the principles stage in specifying a system. To go further would be premature. There is not yet sufficient information to formulate an efficient taxation system for offshore hydrocarbon revenue sharing. But there is a danger that—if revenue sharing is adopted at all, and because of the complexity of the situation and the lack of information—a "simple" system will be chosen: that is, a system which is inefficient for both producers and revenue collectors but which is readily justifiable on pragmatic grounds. It would be far preferable in the preliminary stage to adopt a set of basic principles which lay the foundation for an efficient revenue sharing system once deep-water operations are well under way.

THE FISCAL INCIDENCE OF A REVENUE SHARING TAX

The fiscal incidence of a revenue sharing tax, that is, who ultimately will pay the revenue sharing bill, is not unsurprisingly also a complex issue. The final outcome depends on the form of the tax, government policy on taxation, and the degree of prevailing competition in the market. If the tax pertains to profits, then it cannot readily be passed on to consumers and the full burden will presumably fall upon the firms. If, however, the government in turn gives the firms a tax credit, a part or all of the tax might be paid at the expense of government tax collections. If the government then decides it is not willing to forego these revenues, it may levy an additional tax on its citizens, thereby shifting part or all of the incidence of the tax to the general population.

If output were taxed, then some or all of the tax may be passed on to consumers depending on whether the firm adds the entire tax to its selling price or lowers its selling price to offset all or part of the tax. The question of an offsetting price decrease is related to the elasticity of demand (the proportionate change in quantity relative to the proportionate change in price) for the firm's (or if all firms in the industry are affected by the tax, the industry's) output. The aggregate demand for hydrocarbons is notable for its high degree of inelasticity, and in the short run demand tends to be extremely inelastic—that is, the quantity demanded is hardly affected by a substantial increase in price. Thus an industry facing an inelastic demand function is likely to pass on all or most of any tax on output.

Perhaps the main issue of interest to a lower income country which enters into an agreement with a foreign company or companies to exploit its offshore resources is whether the country, the firms or neither, will be persuaded to accept a smaller share of income as a result of a tax on output. At first glance it would appear that in the exploitation agreement the government would not allow the tax as a production expense, thereby causing the company either to absorb the tax or to increase its selling price. But several factors can be expected to come into play to influence the actual outcome. The first is that, while aggregate demand for petroleum may be inelastic, demand for the firm's output may be elastic because it competes with other firms. If this is the case, adding the tax to the firm's selling price could result in a disproportionately large drop in sales. Thus a company selling its small output in the international market or in a very competitive national market would be forced to bear all or most of the incidence of the tax. If the producing company were selling the petroleum in the same country from which the crude is produced, and if the company exercises some monopoly power, the company would probably be able to recover the tax through a higher price. In this case the host government essentially passes the tax on to the company, which in turn relays it back to consumers in the same country.

Returning to the general case, however, it is conceivable, but not likely, that the industry could be arranged in such a way that firms would bear most of the tax. For example, if the "offshore industry"—producers of crude hydrocarbons—constituted only a small portion of the entire industry in terms of total sales, then it is less likely that the incidence of the tax could be shifted except in instances where offshore producers exercise local monopolies. Finally, if the offshore industry specialized only in the production of crude and sold its output to refiners, even a small price increase would be significant because of the greater attention refiners (as professional purchasers of bulk amounts of crude) pay to supply prices and the discretion refiners exercise in choice of suppliers.

It might be argued that if the international price of crude continues to be determined by OPEC, individual oil companies (or perhaps even companies acting in collusion) cannot charge in excess of the OPEC price. They must absorb the tax if not allowed to charge it as an expense. It is not clear, however, if this would be the case if M.A. Adelman's thesis is correct that OPEC monopoly power is derived from the fact that international oil companies act as its marketing agents; only a small portion of total international crude sales are conducted directly by OPEC members. Thus it is conceivable that companies could add the tax to any OPEC price without causing a proportional shift to direct purchases from OPEC countries.

If it were necessary to make an overall judgment on the ultimate bearer of a tax on output, the structure of the petroleum industry would have to be considered the key determinant in the absence of government policy to direct the incidence. If it is assumed that there will be widespread participation by international oil companies in exploration, and because of their vertical integration and the fact that they tend to dominate international as well as domestic production, processing and sales, the tendency will be to shift the incidence of the tax to final consumers. To the extent that consumers in higher income countries enjoy proportionately more of the benefits from use of petroleum and petroleum products, they can also be expected to pay a relatively greater share of the total tax.

As a closing comment on the fiscal incidence, it should be noted that a revenue sharing tax as a proportion of final product price is likely to be very small. First, the tax is placed on the crude base prior to its passing through several value-added stages. Second, the tax would be spread over the price of output from onshore and close-offshore areas as well as on output from international offshore locations.

REVENUE POTENTIAL AND TAX EFFECTS

As a final step in the analysis of offshore hydrocarbon exploitation with respect to developing economies, some speculative estimates are offered of the approximate magnitude of funds which might be generated from an international community tax on deep-water offshore hydrocarbon production. These figures are clearly very speculative since their derivation involves a large number of variables, few of which can be predicted with any appreciable degree of confidence [31]. The revenue estimation exercise is followed by a simple illustration of the interrelationship between the form of taxation and the actual revenue yield. The illustration also serves to emphasize that meaningful estimation of revenue potential will require much more information than is presently available.

It is possible to make very crude estimates of revenue potential by taking the estimates of the amount of hydrocarbons in given areas, multiplying by some assumed price and taking a percentage of the gross value of production. Thus, at a price of $11.65, the sum of gross revenues from virtually the entire continental margin is equal to $22,836 billion, and a royalty rate of 10 percent yields $228.4 billion. For a more restricted area, for example, the area seaward of the 200-meter isobath, retaining the same price and royalty rate assumptions, the sum of gross revenues would be $84.8 billion. Aside from the obvious assumptions about price and royalty rate, these figures are not very informative because (1) they are not discounted (all of the resources cannot be extracted in one short period), (2) the resource estimates, even assuming they are reasonably accurate, incorporate oil equivalent of the gas in the total estimate, and (3) it is assumed that all of the resources exist in collections sufficiently large to be commercially viable [32].

Some of these problems can be avoided by estimating revenue potential based on annual output as shown below in Table 4–2. Stated in (1974) values for given years there is no need to discount the revenue flow and so long as no statement is made about the duration of production (i.e., the cumulative amount of production relative to the amount ultimately produced), there is no implicit assumption about the size distribution of deposits or the oil-natural gas composition of the resource estimates.

The starting point in estimating annual revenue yields is selection of the revenue sharing area. A few estimates exist as to future off-shore production, but there has been no speculation with respect to given segments of the continental margin aside from a conjecture by McKelvey and Wang that production from the area seaward of the 200-meter isobath in 1968 would be in the range of .5 billion to 1 billion barrels and a projection by the UN Secretary General that in the year 2000 output would reach 2.7 billion barrels of petroleum [33].

The seabed area seaward of 40 nautical miles is adopted as the relevant area for estimation of revenue sharing potential in Table 4–2. There is no particular significance to this specification *per se* other than the fact that it is the largest seabed area, short of including the territorial seabed areas, for which a resource estimate exists. There is a point, however, in choosing the broadest possible area for which resource data are available. The more a tax can be generalized

Table 4–2. Hypothetical Revenue Sharing Potential from Hydrocarbon Production in the Area Seaward from 40 Nautical Miles from Shore

Year	*Output Level*	*Output (10^6 bbls/yr)*	*Value ($\$10^6$)*	*Revenue Sharing Rate (gross royalty payment in millions of dollars)* 1%	5%	10%	50%
1980	High	213.0	2,481	24.8	124.0	248.1	1,240.0
	Low	35.5	413	4.1	20.6	41.3	206.0
1985	High	375.4	4,373	43.7	218.6	43.7	2,186.5
	Low	62.6	729	7.3	36.4	72.9	364.5
1990	High	661.6	7,708	77.0	385.4	770.8	3,854.0
	Low	110.3	1,285	12.8	64.2	128.5	642.5
2000	High	2,054.8	23,938	239.4	1,196.9	2,393.8	11,969.2
	Low	342.6	3,991	39.9	199.6	399.1	1,995.5

with respect to area, the more its potential distortionary effect on location choice is minimized. The potential adverse effect on production efficiency is minimized because producers have less of a scope for choosing sites outside the tax area which are more costly per unit of output, but which are preferred because they provide the producer a higher rate of return. If "all" locations are subject to taxation, the most productive location and the most profitable location become one. Choice of a broad area also makes a forecast of a smooth growth path such as that assumed in Table 4–2 more realistic because output growth for any small area may be quite erratic. To add some perspective, the magnitude of hydrocarbon potential of the 40-mile-and-seaward area is similar to the potential of the area seaward of the 200-meter isobath, 922 billion barrels and 728 billion barrels, respectively. Use of the 40-mile boundary also implicitly incorporates the adoption of a grandfather clause which would exempt ongoing offshore operations.

The revenue generated from a royalty on gross value of output is shown in Table 4–2 to vary in 1980 from a low of $4.13 million to a high of $1.2 billion, depending on the output and royalty levels [34]. By way of comparison, total official development assistance (ODA) to developing nations in 1974 totaled approximately $15 billion. In the same year net resource flows to developing countries were about $40.8 billion [35]. The highest revenue figure shown in Table 4–2 is $11.97 billion, occurring in the year 2000. This revenue figure is slightly higher than total ODA from member countries of the OECD in 1974 [36].

The second part of the revenue estimation exercise is a comparison of two basic types of taxation with respect to their probable effects on output and actual tax receipts. The two taxation forms are a gross revenue tax or royalty such as that used in deriving the revenue figures in Table 4–2 and a tax on net (before tax) profits. The point of the comparison is straightforward: a gross revenue tax—a royalty—will tend to make some offshore operations subcommercial which, under a system of profit taxation, would not be made subcommercial. The result is that a royalty will tend to reduce output below the level which would otherwise occur in the absence of a royalty. Thus any estimate of revenue sharing potential which does not take into account the depressing effect of the royalty tax on production tends to overestimate potential revenue receipts. The relative degree of overestimation or divergence between estimated gross receipts and estimated gross receipts adjusted for the production effect increases the greater the difference in production conditions, and one of the fundamental characteristics of offshore petroleum development is

Table 4–3. Earnings from Offshore Petroleum Production

					Climate	
					Mild	
Water Depth (Meters)	*Price ($/bbl.)*	*Net Annual Earnings ($$10^6$)*	*Recoverable Reserves (10^6 bbl.)*	*Annual Capital Costs ($$10^6$)*	*Annual Net Revenue ($$10^6$)*	*Net Revenue ($/bbl.)*
200	6.25	54.7	50	30.58	24.12	2.64
			100	19.76	34.94	8.83
	8.50	75.2	50	30.58	44.62	4.89
			100	19.76	55.44	6.08
	11.65	103.9	50	30.58	73.32	8.03
			100	19.76	84.14	9.22
500	6.25	54.7	50	89.87	Negative	Negative
			100	58.07	Negative*	Negative*
	8.50	75.2	50	89.87	Negative	Negative
			100	58.07	17.13	1.88
	11.65	103.9	50	89.87	14.03	1.54
			100	58.07	45.83	5.02
1000	6.25	54.7	50	157.71	Negative	Negative
			100	101.93	Negative	Negative
	8.50	75.2	50	157.71	Negative	Negative
			100	101.93	Negative	Negative
	11.65	103.9	50	157.71	Negative	Negative
			100	101.93	1.97	.22

*If the interest rate were reduced from 12 percent to 8 percent, all estimates of net revenue increase appropriately and the negative entries shown with an asterisk become positive.

the existence of very large differences in costs among different production sites.

To demonstrate the diverse effects of the two types of taxation on output and revenue yield, consider the figures in Table 4–3, which is a replication of Table 2–4 except certain figures have been highlighted. In a highly simplified example, assume net revenue in Table 4–3 is equivalent to net profit, and net profit corresponds to economic rent. Further assume the price of oil at the wellhead is $11.65 and the entire seabed area subject to revenue sharing is represented by the eight production complexes highlighted in Table 4–3.

If a 60 percent net profit tax were introduced (thereby allowing

Table 4–3. continued

Climate					
Moderate			*Severe*		
Annual Capital Costs ($10⁶)	*Annual Net Revenue ($10⁶)*	*Net Revenue ($/bbl.)*	*Annual Capital Costs ($10⁶)*	*Annual Net Revenue ($10⁶)*	*Net Revenue ($/bbl.)*
36.88	17.82	1.95	104.35	Negative	Negative
23.83	30.87	3.38	67.43	Negative	Negative
36.88	38.32	4.20	104.35	Negative	Negative
23.83	51.37	5.63	67.43	7.77	.85
36.88	67.02	7.34	104.35	Negative	Negative
23.83	80.07	8.77	67.43	36.47	3.99
107.69	Negative	Negative	Not Available		
69.58	Negative*	Negative*			
107.69	Negative	Negative			
69.58	5.62	.62			
107.69	Negative	Negative			
69.58	34.32	3.76			
188.35	Negative	Negative	409.41	Negative	Negative
121.70	Negative	Negative	264.54	Negative	Negative
188.35	Negative	Negative	409.41	Negative	Negative
121.70	Negative	Negative	264.54	Negative	Negative
188.35	Negative	Negative	409.41	Negative	Negative
121.70	Negative*	Negative*	264.54	Negative	Negative

the coastal state to appropriate 40 percent of the rent element), the total tax yield would be 60 percent of the sum of annual net revenues for the eight complexes, or \$284.44 million (.60 × \$474.06 million). (The figures in this comparative exercise are not comparable in any way to the revenue figures in Table 4–2.) None of the complexes would shut down because all costs, including a return on investment, would be covered after the tax, plus 40 percent of net revenue after taxes. To illustrate the differences between an ad valorem sales tax, or "royalty," and a profits tax, the comparison can be expressed in two ways: the revenue generation capability or the impact on production. While a 60 percent profits tax would yield \$284.44 million annually, it would be impossible to attain this yield using a royalty system. A profit tax revenue yield of \$284.44 million is equivalent to a gross royalty of 33 percent ($284.44 million ÷ total gross revenue, $850.48 million). A 33 percent share of gross revenue per complex (9.125 million barrels at $11.65 each, or $106.3 million) is $35.08 million which, if levied on the eight offshore complexes in Table 4–3, would result in the closing of three complexes because the required tax payment would exceed the annual net revenue (profit) from each of the three complexes. A 33 percent royalty on the remaining five complexes would produce a total tax take of $115.63 million, about 40 percent of the profit tax yield, and society would be deprived of 27.4 million barrels of petroleum production each year. Given the cost and revenue structure of the eight-complex "industry" in Table 4–3, it is impossible to apply a single royalty rate which could yield the same amount of tax revenue as the 60 percent profit tax. If the royalty tax were constrained to a level such that no complex would be made subcommercial by imposition of the tax, then the rate would be only 2 percent (net revenue of the lowest profit complex, $1.97 million ÷ by gross income of $106.3 million), with a revenue generation potential of $8.79 million.

The first and most obvious implication of this comparison is that any estimate of revenue based on a royalty form of taxation which does not take into account the production effect tends to overestimate revenue generation potential. The main thrust of the comparison, however, is to emphasize the appreciable mutuality of interest between producers and any potential recipients of shared revenue. Though highly simplified, the comparison serves to demonstrate that it is clearly in the best interests of both producers and revenue collectors to devise a tax system which does not markedly discourage development. Given the complex nature of offshore hydrocarbon production, a system which can satisfy the requirements of both sides is likely to require a matching degree of sophistication. Otherwise, if an overly simple system is adopted, there is a strong chance it will fail to capture the appropriate share of rent or it will seriously impede development of offshore hydrocarbon resources—or some combination of the two will occur.

THE ALLOCATION OF SHARED REVENUES

The extended Hedberg geo-link thesis serves as a rationale for revenue sharing and implies that all nations, with the possible exception of those states completely surrounded by abyssal plain, have a common claim on resources in the international areas of the continental margins. It has further been argued that the revenue to be shared should be equivalent to the economic rent element contained within the selling price of hydrocarbon resources. These combined points provide a rudimentary prescription for revenue sharing: all states by right of being a part of the geological-geomorphic unit from which hydrocarbons are extracted thereby are entitled to a portion of the economic rent from the sale of hydrocarbons.

Note that if the basic principle is strictly implemented, the revenue sharing obligation goes in both directions; the producing state collects its portion of rents and releases the remainder for distribution. The same producing state would receive a portion of the rents from production in other areas of the geological-geomorphic unit. While consistent with the geo-link thesis, such a system, because of the myriad cross-flows, and even assuming a reasonable degree of cooperation, would tend to be chaotic in operation and, more important, would produce the anomaly of sharing payments being made by poor coastal countries to rich landlocked countries (e.g., by India to Switzerland).

The rather obvious solution to this problem has already been proposed in a number of schemes which would direct the bulk of offshore hydrocarbon revenue sharing from higher income coastal states to lower income economies. The United Nations Committee on the Peaceful Uses of the Sea-Bed has suggested five alternative distribution models. The Committee suggests that per capita gross domestic product (GDP), although it has shortcomings, is probably the best available criterion "for purposes of international comparison" [37]. The five models differ essentially in the weight each assigns to the level of GDP, which in turn indicates the amount of payment each country receives. The use of per capita GDP as a distribution criterion is not in rigorous adherence to the geo-link principle (which if narrowly defined would require distributing the revenue in accordance with some physical criterion such as the proportional area of the recipient country) but, given the distribution of offshore potential relative to national per capita income, GDP per capita is not a bad proxy.

The first UN plan would distribute funds inversely with the country's relative share in world Gross Domestic Product and in positive relation to the country's share of world population. In this scheme relative GDP ranking and population ranking would have equal weight in the allocating formula. This plan obviously favors those countries with extremely large populations and fairly low GNP level. Many of the least developed countries which are also landlocked or shelf-locked have relatively small populations and are thus not well served by this plan. It should be pointed out that the UN plan applies to all countries: that is, developing and developed countries would all share in this pool of resources. It is, therefore, more or less consistent with the implications of the extended Hedberg thesis.

The second UN plan would divide the pool of funds into three equal blocks, with the first block being shared by all countries irrespective of per capita GDP, the second block being shared by all countries with per capita GDP below $1000, and the third block being shared by those countries with per capita GDP below $150. Distribution within each block would be determined by each country's relative population share.

The third UN variant would establish five blocks with an allocation as follows: (1) to all countries; (2) to countries with per capita GDP under $1000; (3) to countries with per capita GDP under $500; (4) to countries with per capita GDP under $250; and (5) to countries with per capita income under $100. Again, shares within blocks are allocated according to population.

In the fourth variation, the shares of countries with per capita GDPs above $500 would be reduced, with the sum of the reduction being redistributed to countries below the $500 level. For countries above $500, the retained share would be determined by each country's share of population divided by the ratio of per capita GDP to $500. The redistributed portion would be divided into two equal parts, the first being distributed proportionately to population to all countries below the $500 mark and the second part being concentrated on the least developed economies.

In the fifth model, the shares of countries above $500 would be reduced in the same way as in model four. The revenue would be redistributed in a progressive fashion and in an inverse relationship to per capita income.

An entirely different approach to the distribution of benefits would involve the transmission of revenue-sharing funds to a development finance institution(s) for use in funding additional development projects. This approach would essentially ignore the basic geo-link relationale for distribution. Instead, revenue-sharing funds would be added to the pool of development assistance funds and disbursed on the basis of the lending criteria of the particular institution (e.g., the International Development Association (IDA), the International Bank for Reconstruction and Development (IBRD), or regional development banks such as the Asian Development Bank or the Inter-American Development Bank).

The idea of distribution through international financial institutions introduces a number of points worth considering. First, an implicit intercountry distribution decision is made in the act of choosing which development lending institutions will receive sharing funds and the amount each institution will receive. If regional institutions are used, then the question of interregional distribution arises. If revenue-sharing funds are transferred to a multilateral institution, then another kind of distribution decision is made, because the multilateral institutions such as the IBRD or the IDA have particular lending criteria and therefore a particular set of clientele countries.

Second, if the funds are disbursed according to some conventional lending criteria which reflects the respective social marginal productivity of each project, regardless of the particular lending institution, the flow of funds would tend to favor those countries within the clientele group which can produce the largest number of projects with the greatest social marginal product [38]. Even if it is assumed that funds are distributed to a number of development finance institutions, distribution to countries within each clientele group would tend to favor the most dynamic and richly endowed economies because they tend to have the most economically and financially attractive projects. One of the general characteristics of the very poorest economies is that they have a very limited capacity to generate promising development projects. If this bias were recognized and an attempt were made to insure that all countries receive a portion of the sharing funds, then the lending criteria of the development institutions would be compromised.

A third point is that to the extent that disbursal of revenue-sharing funds by international development institutions are loans (allowing for a varying grant element), the loan element constitutes a bias against the recipient countries. Producing, coastal states which acquire their share of hydrocarbon revenues receive them without obligation of repayment. Countries which receive their share of hydrocarbon revenues as loans (or partially as loans) have to relinquish the loan component at some future date. In addition, the grant element of revenue-sharing funds is financially equivalent to the rent received by producing states, but the grant portion still differs in that its receipt is tied to the decision of the particular development institution, whereas, receipt of funds by the producing coastal state would be automatic and unencumbered.

As a last point, the distribution of revenue-sharing funds again introduces the additionality problem noted earlier on page 100. International financial institutions are supported by appropriations from national governments (basically, the appropriations go to offset the grant element in each institution's lending). If revenue-sharing funds were to be channeled through international development institutions, there is a strong probability, in view of recent attitudes expressed by some donor countries, that the increase would encourage national governments to reduce their levels of participation, thus partially or completely offsetting the incremental increase in development assistance. In conclusion, channeling revenue-sharing funds through international financial institutions may increase the efficiency in the use of the funds but it also may introduce a bias against their equitable distribution among developing countries and it may result in offsetting reductions in existing development assistance from donor nations.

It can readily be seen that there is potential for a large number of alternative schemes for distribution of benefits among developing nations. Although economic analysis can be of use in the decision process, the ultimate distribution decision is largely subjective. It seems appropriate, therefore, that the question of benefit distribution among developing countries be settled in the fora of the United Nations with the low-income countries playing a major and informed role.

OFFSHORE HYDROCARBONS: SUMMARY AND CONCLUSION

This chapter on the disposition of benefits from offshore hydrocarbon production has stressed (1) the rapid reversal since 1970 from internationalization to nationalization of seabed hydrocarbons, (2) the existence of economic rent in seabed hydrocarbons, (3) the existence of a substantive rationale for a claim by geographically disadvantaged countries to a portion of the rent element in seabed hydrocarbons, and (4) the fact that revenue sharing, even assuming widespread acceptance, is a genuinely complex issue but at the same time one in which there is a substantial area of potential mutual interest (maximization of revenue within agreed limits and minimum hindrance to production).

In spite of the potential magnitude of wealth in ocean hydrocarbon resources, simplistic and probably optimistic estimation of revenue sharing magnitudes indicates that the flow of official development assistance is likely to remain the predominant source for financing economic development. The generation of revenue-sharing funds is sensitive to the manner in which they are collected, which suggests that adoption of a "pragmatically simple tax" could be detrimental to both producers' and collectors' interests. As a result, it is suggested that the initial step in the formulation of a revenue sharing scheme consist of an agreement on broad principles which prescribe a sharing system with a degree of economic sophistication comparable to the degree of economic complexity which characterizes offshore hydrocarbon production. The development of deepwater seabed hydrocarbon resources, over time, would then provide the information base needed for setting out a detailed revenue sharing system.

With regard to the fundamental issue of the attitude taken by coastal states toward revenue sharing for the benefit of low-income countries, it is noted that 1976 marks the 482nd anniversary of the Treaty of Tordesillas [39]. It was in this treaty that Spain and Portugal agreed on the division of the New World between themselves. Now the time has come for the nations of the world in the Law of the Sea negotiations to agree on the division of the new world of offshore hydrocarbon resources. The outcome will provide a clear indication of the direction and degree of evolution in international economic and political affairs since Spain and Portugal last divided the world in 1492.

NOTES TO CHAPTER 4

1. Evan Luard, *The Control of the Sea-bed: A New International Issue* (New York: Taplinger Publishing Co., Inc., 1974), p. 84.

2. Argentina, Brazil, Ecuador, El Salvador, Panama, Peru, Uruguay, Sierra Leone and Somalia already claim a 200-mile territorial sea. Colombia is in the process of extending its claim to 200 miles, and Mexico claims a 200-mile exclusive economic zone. (U.S. Department of State, Bureau of Intelligence and Research, Office of the Geographer, *Limits in the Seas no. 36: National Claims to Maritime Jurisdictions*, 3d rev. (Washington, D.C., 23 December 1975).

3. U.S. Department of State, "The Third United Nations Conference on the Law of the Sea: US Delegation Report," Caracas, Venezuela, 20 June–29 August [1975]. (Mimeographed.)

4. *Ibid.*, p. 16.

5. The idea that economic rent is basic to exploitation of the seabed and its subsoil is evident in the 1970 United States document, "Draft United Nations Convention on the International Seabed Area: Working Paper" (Washington, D.C., 20 August 1970, p. A 12). Richard Cooper also discusses the rent element of ocean floor minerals and proposes a scheme to capture the bulk of economic rents ("An Economist's View of the Oceans," in *Perspectives on Ocean Policy: Conference on Conflicts and Order in Ocean Relations, Arlie, Virginia, 21–24 October 1974*, prepared for the National Science Foundation by The Johns Hopkins University Ocean Policy Project, Washington, D.C., pp. 152–53).

6. Economists sometimes call the return on investment "normal profits"—the level of profit just sufficient to stay in business in the long run. Normal profit is a cost of providing the product or service.

7. A more detailed discussion of economic rent as it applies to crude oil

would divide the rent component itself into two components, scarcity rent and monopoly rent. Scarcity rent is the difference between the total long-run production cost and a competitively determined selling price. Monopoly rent (frequently called quasi-monopoly rent) is the difference between a competitively determined price and a higher price which is obtained by restricting output. (M.A. Adelman, "The World Oil Outlook," in *Natural Resources and International Development*, ed. Marion Clawson [Baltimore: The Johns Hopkins Press for Resources for the Future, Inc., 1974], pp. 27–113).

8. If the rents from offshore production could be calculated and aggregated and if the proportionate sharing weight for each country were known, it would be possible to determine the value of the rents from which the geographically disadvantaged states were excluded. The loss to these states would be equal to total rent per period divided by the summed weights of producing countries less aggregated rent per time period divided by the sum of the sharing weights of all countries,

Let R_j be defined as the economic rent derived from crude oil extraction off the coast of the jth country, where R_j is defined for $j = 1 \ldots n$ coastal states. In the absence of revenue sharing, each jth state receives R_j amount of economic rent ($j = 1 \ldots n$). Non-coastal states, $j = n + 1 \ldots m$, receive nothing.

With a system of revenue sharing, each country is assigned a sharing weight, W_j, and receives

$$\frac{W_j}{\sum_{j=1}^{m} W_j} \left(\sum_{j=1}^{n} R_j \right) .$$

Coastal (non-coastal) states in the aggregate stand to lose (gain) an amount of economic rent equal to

$$\sum_{j=n+1}^{m} \left(\frac{W_j}{\sum_{j=1}^{m} W_j} \left[\sum_{j=1}^{n} R_j \right] \right) .$$

An individual jth coastal state would lose the following amount, assuming that the total exploitation (and rent) available is the same with or without revenue sharing,

$$R_j - \frac{W_j}{\sum_{j=1}^{m} W_j} \left(\sum_{j=1}^{n} R_j \right) .$$

9. Hollis Hedberg is professor emeritus of geology at Princeton University and was formerly a vice-president for exploration for the Gulf Oil Corporation (Hollis D. Hedberg, "The National-International Jurisdictional Boundary on the Ocean Floor," *Ocean Management* 1 [1973]:84).

10. *Ibid.*, pp. 86, 90.
11. *Ibid.*, pp. 87, 91.
12. *Ibid.*, p. 88.
13. *Ibid.*, pp. 89–90.
14. Hedberg alludes to this point in a discussion with participants at the Fifth Annual Conference of the Law of the Sea Institute (Lewis M. Alexander, ed., *The Law of the Sea: The United Nations and Ocean Management, Proceedings of the Fifth Annual Conference of the Law of the Sea Institute, June 15–June 19, 1970* [Kingston: The University of Rhode Island, January 1971], p. 164).
15. Perhaps claim allocation could be carried out by some formula which, for example, takes into account the area of each country of a continent and perhaps other factors such as population size and income per capita. As an alternative, allocation could conceivably be based on a random drawing, e.g., country A gets parcel 23, country B gets parcel 16, and so on.
16. U.S. Department of State, Office of the Geographer, "Major Topographic Divisions of the Continental Margins" (May), July 1970. There are some states which are surrounded by seabeds at water depths greater than 200 meters, e.g., New Zealand, Iceland, Maldives and the Malagasy Republic, but they are not surrounded by abyssal plain and therefore remain part of the globe's positive element.
17. While the details differ, the principles of the debate are similar to those in the long-standing struggle between the United States federal government and a number of state governments. The attempts of some coastal states in the United States to extend their authority over adjacent offshore petroleum resources have been opposed by the federal government, and in a series of court decisions state jurisdiction has been limited generally to the area within 3 miles from shore, while the federal government (and, implicitly, states without offshore hydrocarbon resources) have retained the resource rights to the United States outer continental shelf.
18. A comprehensive but terse presentation of the benefits and ability-to-pay principles is provided in Richard Musgrave, *The Theory of Public Finance* (New York: McGraw-Hill Book Company, 1959), Chaps. 4 and 5. See also pp. 61–63 for a succinct discussion of the two principles.
19. See K. Lancaster and R. Lipsey, "The Theory of Second Best," *Review of Economic Studies* 24, no. 63 (1956–57):11–32. See also pp. 207–208.
20. Hotelling, "Economics of Exhaustible Resources." For a recent survey of the topic, including a digest of the Hotelling article, see Solow, "Economics of Resources."
21. There is no doubt that coastal states in general would object to a coastward boundary of 12 nautical miles, and states which already have production sites beyond 12 nautical miles could be expected to be particularly vociferous in their opposition (e.g., riparian North Sea countries, the United States). A relatively routine compromise solution exists for these situations, however, in the form of a "grandfather clause." This clause would exempt from revenue sharing, producing sites which were in operation prior to enactment of the revenue measure. Obviously, there are a number of variations on this theme, a fact which makes it a potentially useful negotiating tool for all parties.

22. In practice the pricing of crude oil is more complex because there are many different kinds of crude; however, crude oils are standardized by their chemical properties and prices are adjusted appropriately. Although minor relative to many of the other aspects confronting the designers of a revenue-sharing system, the use of existing reference prices would require caution because some crude oil prices include a location premium or discount (e.g., Algerian crude of a given grade may have a higher wellhead price than the same grade of crude produced at a location further away from the European market). Thus any reference price chosen must be neutral with respect to location or, if a price could not be found which did not include a transportation premium or discount, then a base point price would have to be chosen.

23. It might be added that while the producing states are likely to object strongly to a profits tax, it is not unique in this respect. The "objection of coastal states" criticism is likely to be applicable to any proposal which would have the effect of reducing producing state gains from offshore oil production.

24. In the case of oil it is not clear whether or not marginal costs (the cost of each additional unit of output) increase with increasing output prior to the secondary recovery stage because marginal costs may remain relatively constant so long as the output rate is not forced to exceed the flow characteristics of the particular deposit. It is conceivable that increasing the output level to the point where marginal cost just equals marginal revenues is not a realistic option. If this speculation were valid, however, the profits tax would still retain its advantage.

25. R. Garnaut and A. Clunies Ross, "Uncertainty, Risk Aversion and the Taxing of Natural Resource Projects," *The Economic Journal* 85 (June 1975): 272–87.

26. *Ibid.*, p. 275.

27. *Ibid.*

28. Cooper, "An Economist's View," p. 153.

29. Kanenas, "Wide Limits and 'Equitable' Distribution of Seabed Resources," *Ocean Development and International Law Journal* 1, no. 2 (1973):137–58.

30. United States proposal in Committee One of the United Nations Third Law of the Sea Conference in Geneva, April–May, 1975.

31. Jonathan I. Charney, "The Equitable Sharing of Revenues from Seabed Mining," *Public Issues in Ocean Law: Studies of the Working Group on Technical Issues of the Law of the Sea* (St. Paul, Minn.: West Publishing Company for the American Society of International Law, 1975), pp. 53–120.

32. Inclusion of the gas equivalent (6000 cubic feet equal 1 barrel) overestimates revenue potential because deep-water offshore gas production is more costly than comparable petroleum production and the price of 6000 cubic feet of natural gas is appreciably less than that of one barrel of oil. Therefore production of gas is likely to be more restricted than oil production.

33. U.S. Department of Interior, *World Subsea Mineral Resources*, by V.E. McKelvey and F.F.H. Wang (Washington, D.C.: Government Printing Office, 1969), p. 9; and United Nations, General Assembly, Committee on the Peaceful Uses of the Sea-Bed and the Ocean Floor beyond the Limits of National Jurisdiction, *Possible Impact of Sea-Bed Mineral Production in the Area beyond National Jurisdiction on World Markets, with Special Reference to the Problems*

of Developing Countries: A Preliminary Assessment, Report of the Secretary General (A/AC.138/36), 28 May 1971, p. 43.

34. The annual output figures in Table 4–2 are derived in the following fashion. Table 1–4 indicates that the ultimate production potential in the continental margin seaward of 40 nautical miles from shore is 922 billion barrels of oil and oil equivalent, of which 284 billion barrels are located seaward of 200 miles from the coast. For Table 4–2 this latter figure is reduced by 75 percent in accord with L.G. Weeks' speculation that less than 25 percent of the petroleum deposits may exist in collections sufficiently large for commercial recovery. The remaining 709 billion barrels is arbitrarily reduced by 50 percent, also to allow for subcommercial deposits. The remaining figure of 335 billion barrels is not inconsistent with the speculation in *World Oil* that the world's *entire* offshore area may ultimately yield 700 billion barrels of oil, keeping in mind that the margin area within 40 nautical miles from shore includes all of the present major offshore producing areas in the world including the Persian Gulf and major portions of the Gulf of Mexico and the North Sea ("Increasing Demand for Oil Spurs Offshore Growth Rate," *World Oil* 177 [July 1973]:82). To derive the high and low estimates for 1980, it is assumed that as a low estimate .01 percent of a 355-billion-barrel reserve will be produced, but a sixfold upward variation in productivity is allowed for to take into account the very large differences in productivity which can be encountered in oil extraction. The productivity range is guided by the sixfold difference in average output rates among the five largest oilfields in the Gulf of Mexico, at a 62,000-barrel-per-day rate, and the average output rate of the five largest fields in the North Sea, with a 360,000-barrel-per-day rate. (Irvin L. White *et al.*, *North Sea Oil and Gas: Implications for Future United States Development* [Norman: University of Oklahoma Press for the Council on Environmental Quality, 1973]. Table 4, p. 55). Thus the initial high option is .06 percent of total reserves. The low option of 35.5 million barrels output in 1980 is equivalent to approximately four of the 2-platform, 40-well complexes described in Chapter 2. The high option of 213 million barrels in 1980 is comparable to the output from approximately twenty-three of these complexes. While it is obviously not possible to make a concrete statement about the amount of petroleum which will be produced from the 40-mile-plus area, even a level of 35.5 million barrels seems optimistic because it would require (assuming the existence of a grandfather clause) the initiation and completion of a substantial production complex(s) within three years, an unlikely event. In other words, the output estimates, at least for the early phase 1985–1990, are probably optimistic.

The output estimates for 1980 are considerably less than those of McKelvey and Wang, but the estimates in Table 4–2 have the advantage of having been made seven years later. Of course, to make any kind of positive estimate of deep-water oil production carries with it the implicit assumption that offshore development will not be appreciably impeded by institutional factors or by uncertainty about future institutional arrangements. To move into the future from 1980, output is assumed to grow at a 12 percent annual rate. It is difficult to evaluate the prospective growth rate except to say that the rate of increase in offshore oil production has tended to exceed the rate of increase in total production for the world. For example, world output increased by 6.6 percent between 1970 and 1971 (growth averaged 7.7 percent annually between 1963 and 1973), whereas world offshore production increased by 16 percent between 1970 and 1971 (The British Petroleum Company Limited, *BP Statistical Review of the World Oil Industry, 1973*, [London: Britannic House, 1975], p. 19; *Offshore* 32 [20 June 1972]:81). A slightly lower rate is adopted to reflect the fact that the area under discussion is, other things being equal, a low priority area relative to the less costly 40-mile band of inshore seabed and therefore subject to a slower growth rate than the close inshore area.

35. Organization for Economic Cooperation and Development, *Development Co-operation: Efforts and Policies of the Members of the Development Advisory Committee*, [Paris], November 1975. Official development assistance consists of "flows from the official sector that are concessional in character—grants, and loans with a grant element of at least 25 percent—and which are undertaken primarily for the purpose of development" (pp. 14–15).

36. *Ibid.*

37. United Nations General Assembly, Committee on the Peaceful Uses of the Sea-Bed and the Ocean Floor beyond the Limits of National Jurisdiction. *Possible Methods and Criteria for the Sharing by the International Community of Proceeds and Other Benefits Derived from the Exploitation of the Resources of the Area beyond the Limits of National Jurisdiction: Report of the Secretary General* (A/AC.137/38), 15 June 1971, pp. 14–22.

38. See O. Eckstein, "Investment Criteria for Economic Development and the Theory of Intertemporal Welfare Economics," *Quarterly Journal of Economics* 71 (February 1957):56–85; W. Galenson and H. Leibenstein, "Investment Criteria, Productivity, and Economic Development," *Quarterly Journal of Economics* 69 (August 1955):343–70; A. Kahn, "Investment Criteria in Development Programs," *Quarterly Journal of Economics* 65 (February 1951): 38–61.

39. Federico G. Gil, *Latin American–United States Relations* (New York: Harcourt Brace Jovanovich, Inc., 1971), pp. 15–16.

Chapter 5

The Nature of Deep Seabed Mineral Resources: Manganese Nodules

INTRODUCTION

The world's oceans are enormous untapped repositories for a large number of minerals. Aside from hydrocarbons located in the submerged portions of the continents, greatest interest has centered on manganese nodules—deep seabed mineral deposits containing nickel, copper, cobalt, and manganese—which lie virtually uncovered on the ocean floor, anywhere from 1000 feet to 20,000 feet beneath the ocean surface. Nodules are unique insofar as they are ubiquitous to the deep seabed and are generally located, at least in concentrations which are attractive for commercial mining, in international ocean areas in which no state has clearly defined rights of (or prohibitions against) mineral exploitation. Although the existence of manganese nodules has been known since 1876, when the British oceanographic research vessel HMS Challenger surveyed the Pacific, it is only in the past decade that advances in ocean mining technology have catapulted nodule mining into the realm of commercial possibility, and concomitantly placed seabed mining rights into the fulcrum of the international negotiations in the Law of the Sea Conference.

Developing countries are concerned that international seabed resources not be appropriated by the rich nations able to develop the sophisticated and expensive technology to exploit mineral deposits on the ocean floor. Concern about seabed mining is particularly intense on the part of developing country land-based producers of seabed minerals, whose foreign exchange earnings are heavily dependent on mineral exports, and who are concerned that seabed production will adversely affect minerals prices and thus lower earnings. This is one of the reasons why developing countries are opposed to unencumbered deep seabed mining competition, preferring a system whereby the international community—with LDCs maintaining a large voice in decisions—determines the appropriate level of manganese nodule mining.

One of the peculiar characteristics of manganese nodules is that, because of the joint nature of production of the contained minerals, the relative amounts of seabed minerals produced will not parallel world demand for these minerals. For example, were seabed mining to supply 50 percent of current world nickel demand, it would probably supply no more than 3.3 percent of world copper demand, but 28.4 percent of world manganese ore demand and approximately 216 percent of world cobalt demand. It is therefore in the economic interest of some mineral exporters, for example, cobalt producers, to attempt to keep deep seabed mining at a minimum.

The issue of who may exploit deep seabed mineral resources, and under what conditions, has been the major focus of the negotiations concerning an international deep seabed mining regime in the recent Law of the Sea deliberations. Industrialized nations, major mineral consumers and potential deep seabed miners, are eager to exploit a new, low-cost source of supply for important minerals and thus wish to gain unimpeded access to deep seabed resources. Developing nations, who are minor mineral consumers and are unlikely to participate in deep seabed mining, desire to prevent the industrialized nations from unilaterally exploiting resources often referred to as the "common heritage of mankind" [1]. By common heritage is implied that the deep seabed resources, lying predominately on the ocean floor outside the limits of national coastal state resource jurisdiction, are not the appropriable property of any one state, but rather are resources to be used for the benefit of mankind. Exactly how such a grand precept is to be rendered operational has proven to be a challenge rivalling that confronting Sisyphus. The modern Hades is Committee One of the UN Law of the Sea Conference.

Our examination of deep seabed resource exploitation and the economic interests of developing nations is divided into three parts. In the remainder of this chapter, we examine the geological characteristics of manganese nodules, the deep seabed mining technology which is being developed to mine nodules commercially by 1980, and the interests of particular developing country land-based producers of seabed competitive minerals. In Chapter 6, we examine the financial viability of nodule mining, including projections of future minerals prices in a post-seabed mining milieu, estimated mining costs, and profits associated with nodule mining investments. We also examine the effects of deep seabed mining on LDC mineral producers to gauge the extent to which some form of adjustment assistance might be useful for LDC mineral producers adversely affected by nodule mining. Chapter 7 contains an examination of the entire range of issues involved in seabed mining; namely, who may exploit the deep seabed and under what conditions? What are the economic implications of positions espoused by countries in the Law of the Sea Conference? What rationale exists for international taxation of deep seabed mining, and what types of taxes are feasible? How might such a tax be implemented, and what are its effects on economic efficiency? In analyzing these issues we will be guided by a dual concern, for efficiency and equity. To exploit an economic resource efficiently is to produce until the cost to society of producing an additional unit of output just equals the benefit society reaps from an additional unit of output. Only by producing efficiently can society allocate its scarce resources in an optimal way. Since ocean resources outside the limits of national jurisdiction are common to all of mankind, an equitable system of distribution of gains is also necessary if all countries are to participate in the benefits to be derived from seabed mining. The world's poorer nations, which are not expected to be directly involved in deep seabed mining, should still share in the benefits.

THE GEOLOGY OF MANGANESE NODULES [2]

While offshore mining has to date concentrated on hydrocarbons and other minerals found in the relatively shallow waters of the continental margins, a handful of mining companies from the industrialized states have spent almost $200 million to develop a new deep ocean mining technology to obtain nodules from the deep seabed and the chemical process needed to extract specific metals [3]. A decision to exploit a mineral deposit—be it a deposit of offshore hydrocarbons, offshore coal or diamonds, manganese nodules or a land-based deposit—is dependent on the geological nature of the find, the availability of the requisite technology and the economic feasibility of the investment. Recent investigations have shown that there are a large number of nodule deposits of commercial interest, deposits of high nodule concentration (i.e., nodules per area) and of highly desirable mineral content. While all nodules contain copper, nickel, cobalt and manganese, only particular deposits contain these minerals in quantities sufficient to make them commercially exploitable at current world prices. And among these mining deposits some are referred to as "prime sites," in part because the quantities of minerals contained in the nodules are such that their exploitation is most profitable [4].

Geologically speaking, manganese nodules are one form of black, hydrous, manganese-oxide agglomerations found in the pelagic (i.e., deep water) areas of the oceans. Manganese-oxide is found in grains as part of pelagic sediments called red clay, in biogenic oozes, as rock coatings, on sharks' teeth and whale bones, and as slabs and nodules. While existing in a plethora of shapes, nodules are most generally semi-spherical or potato-shaped. Nodules, varying in diameter from less than one mm. to 25 cm., are formed by the accretion of charged manganese and iron-oxide ions and other materials, aided in some measure by deep sea currents. Scientific opinion differs concerning the rate of formation of nodules—they are reported to increase in size through accretion at a rate between 1 mm. per 100 years and 1 mm. per 100,000 years [5]. Assuming an accumulation rate of 1 mm. per 1000 years, it would take 200,000 years to replace the existing stock of nodules in the Pacific Ocean alone.

Through sampling and photographic surveillance techniques, considerable knowledge has been gained on the location, concentration and mineral content of nodules. There are few places in the ocean except the continental shelves and deep ocean trenches where nodules cannot be found, and many locations contain significant concentrations of nodules. Figure 5–1 shows the worldwide distribution of nodule deposits. Estimates of manganese nodule concentrations vary from less than 1 gram per cm^2 to approximately 1.5 grams per cm^2, which corresponds to a range of perhaps 28,000 to 42,050 short tons of nodules per square mile [6]. An estimate for the total quantity of manganese nodules lying on the Pacific seabed alone is 1.66 trillion metric tons [7].

The mineral content of nodules varies widely among alternative mining areas and to some extent within areas [8]. According to Mero, geologists have identified four basic types of nodules: (1) high iron content nodules in which the manganese to iron ratio is below unity and which contain approximately 28 percent iron, 22 percent manganese and less than 1 percent of both nickel, cobalt and copper; (2) high manganese content nodules with average assays of close to 50 percent manganese and small amounts of nickel, copper, cobalt and lead; (3) high cobalt content nodules with average assays of 29 percent manganese, 23 percent iron, 1.2 percent cobalt, and small

Figure 5–1. The Worldwide Distribution of Manganese Nodule Deposits

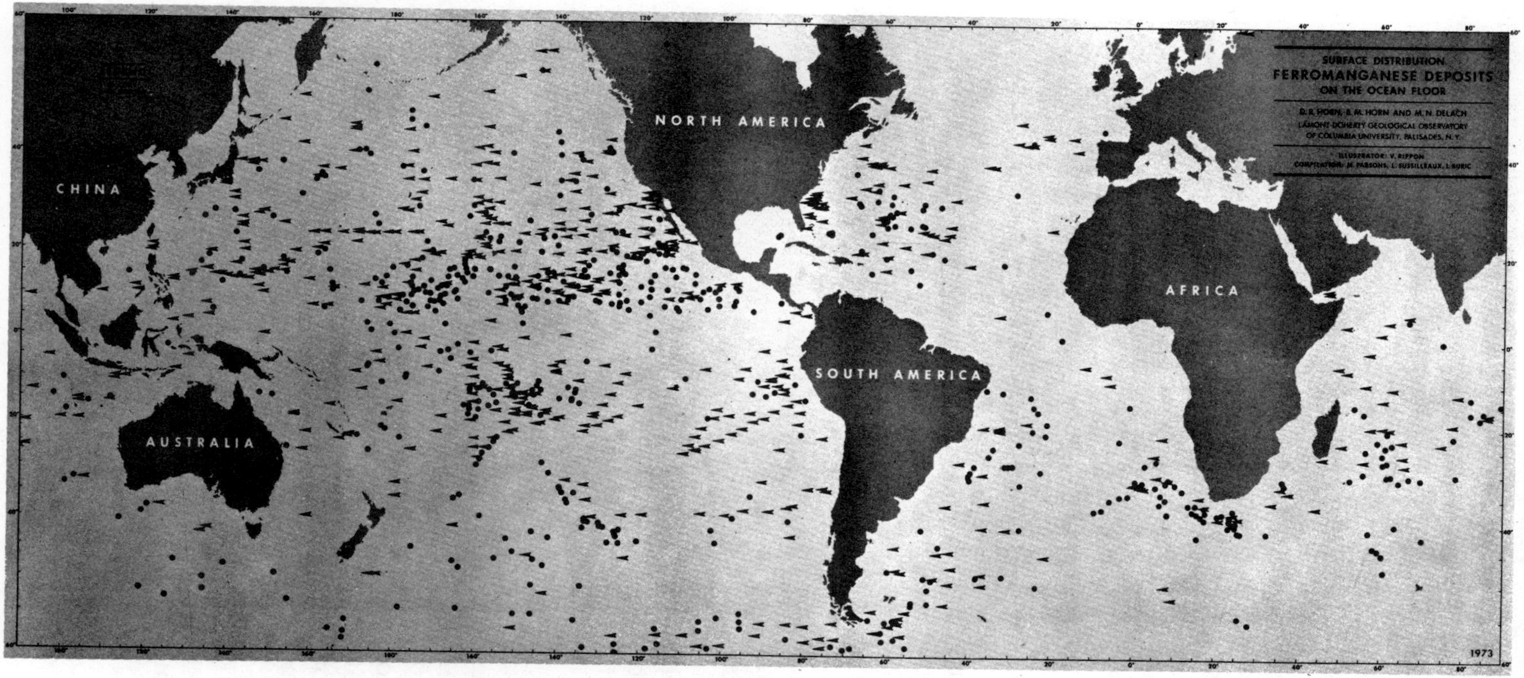

Source: D. Horn, B. Horn, and M. Delach, "Ocean Manganese Nodules—Metal Values and Mining Sites," Technical Report No. 4, National Science Foundation, 1973, reprinted with permission from the N.S.F.

amounts of nickel, copper and lead; and (4) high nickel and copper content nodules with average assays of 33 percent manganese, 18 percent iron, 1.5 percent nickel, 1.1 percent copper, .4 percent cobalt and .2 percent lead [9]. It is this last category of nodules, rich in both copper and nickel, that has attracted mining companies' interest, since it is the high nickel-copper content that renders a nodule deposit profitable to exploit at current world prices. A general rule of thumb is that a deposit should contain at least 1 percent each of nickel and copper to be economically viable under current economic conditions.

A recent geological report indicates that nodules containing high percentages of nickel and copper are predominant in areas of the North Pacific on an east-west band south of Hawaii and north of the equator from approximately 2000 miles west of Costa Rica westward to the Marshall Islands [10]. Figure 5–2 depicts those seabed areas with the largest concentrations of high nickel content nodules. Indicative of the interest of mining concerns in this rich mineral area is the claim filed by Deepsea Ventures, an American deep ocean mining company formed expressly to exploit manganese nodules as early as 1980, to a prime area of 60,000 square kilometers approximately 500 miles south and 2500 miles east of Honolulu [11]. The mining claim was filed November 15, 1974 and a request for diplomatic protection sent to the Department of State [12]. The Deepsea Ventures Ocean Mining Claim reported that the average water depth in the claim area is between 7500 and 16,400 feet and that the average estimated concentration of nodules is equivalent to 2.15 lbs. per square foot of dry nodule material [13]. An average assay as a percent of dry weight is reported to be 1.28 percent nickel, 1.07 percent copper, 0.25 percent cobalt, 29 percent manganese and 6.3 percent iron.

Deepsea Ventures, along with at least two other active consortia, is developing the technology to be ready to mine manganese nodules at depths of 10,000 to 20,000 feet by the end of this decade. Firms are making these investments in research and development, accepting the risks associated with both uncertain future mineral prices and a high degree of uncertainty about the kind of system which might eventually govern nodule exploitation. The international community, on the other hand, is being asked to negotiate a system of deep seabed mining rights in a similar cloud of uncertainty. A useful point of departure in an analysis of the economic interests of developing nations in nodule mining is to consider the characteristics of the new nodule mining technology.

Figure 5–2. Worldwide Distribution of Manganese Nodules by Nickel Content

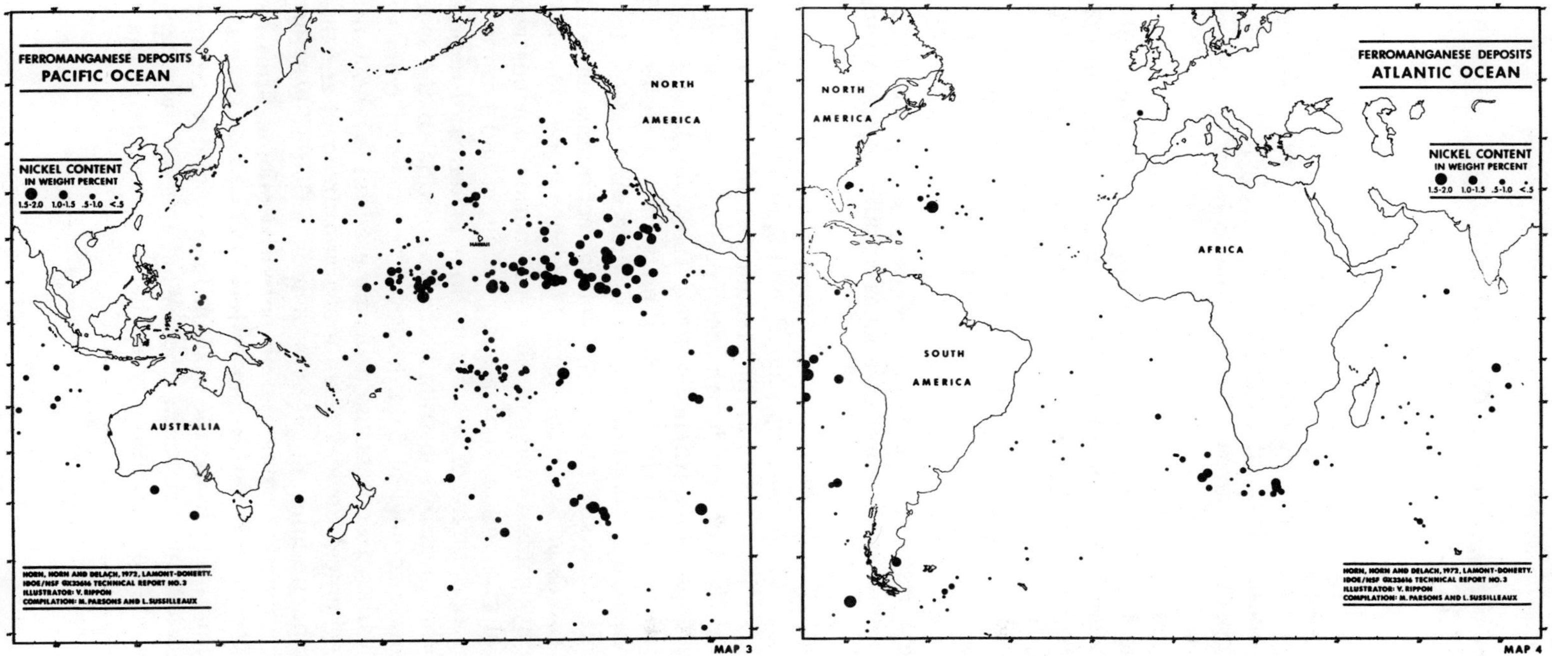

Source: Horn, Horn and Delach, *op. cit.*, with permission from the National Science Foundation.

THE TECHNOLOGY OF MINERAL EXPLOITATION [14]

The technology of deep seabed manganese nodule mining can be separated into four basic activities: (1) surveying and prospecting; (2) gathering and lifting the nodules to the ocean's surface; (3) transporting the nodules to a processing area; and (4) chemically processing the nodule material to extract the contained metals. The total mining operation can be characterized as being highly capital-intensive and technically complex.

A review of the pattern of investment in deep seabed mining confirms the fact that only the industralized countries will be able to pursue these kind of operations in the foreseeable future. Firms committed to manganese nodule mining projects are Deepsea Ventures, a consortium involving Tenneco, U.S. Steel, Union Minière of Belgium, and a number of Japanese companies; a Kennecott consortium of Kennecott Copper, Rio-Tinto Zinc, Consolidated Gold Fields, Noranda Mines and Mitsubishi; an International Nickel Company of Canada (INCO) consortium including the German AMR group, SEDCO and a Sumitomo-led Japanese group (DOMCO); and possibly Summa Corporation (although recent accounts concerning the deep sea salvaging activities of the Summa Corporation mining ship place the role of Summa in deep seabed mining in doubt). Thus, the development of a deep seabed mining technology involves companies from only a handful of the industrialized countries, a fact which is of considerable importance when discussing the attitude of the developing nations towards the exploitation of manganese nodules [15].

Surveying and prospecting activities are required for the selection of a prime mine site [16]. Prospecting techniques have become extremely sophisticated, with optical searches using closed-circuit TV, and geophysical tests using sensitive magnetic and acoustical systems. Acoustical systems, including sonar scanning and seismic tests, help map the topography of the ocean floor in terms of depth from the surface, thickness of substrata and composition of the seabed itself. The ultimate geological data, however, are derived from actual sampling. A variety of corers and drills are used for sub-bottom sampling, while mechanical devices (e.g., grabs and dredges) are employed to recover the nodules themselves. These activities generally require special prospecting ships and all potential mining firms have undertaken research and prospecting cruises [17].

Mechanical and hydraulic systems have been devised to gather and lift nodules from the ocean floor. What renders these technological achievements so revolutionary is that they enable recovery of nodules lying under 2 to 3 miles of water and the operations can be carried out in a variety of ocean conditions. The only mechanical system with apparent commercial interest is the continous line bucket (CLB) system which involves a series of dredge buckets hung by cable from the surface vessel or vessels. The cable is submerged in a loop fashion and is continuously rotated with successive buckets digging, being filled, moving forward and being lifted to the surface for dumping. Although the flexible cable is fairly adaptable to currents and variable topography, potential material loss at significant depths and in heavy currents is said to exceed that of hydraulic methods. CNEXO of France and Sumitomo of Japan are reported to be planning the testing of CLB systems at water depths as great as 20,000 feet; this would require at least 8 miles of cable.

Hydraulic methods of recovering nodules represent a more sophisticated technology. The airlift pump system being developed by Deepsea Ventures involves the pumping of air into a long (i.e., 15,000-foot) flexible pipe attached by a shorter, rigid pipe to the dredgehead, which is dragged along the ocean floor. Lift is created when the injected air expands in the pipe and forces the lighter water-air mixture to rise. This suction system forces both water and the surficial deposits of nodules to the surface without significant sediment disturbance. The material is forced upwards, outflows are collected and separated at the surface, and new material rushes in at the pipe base. The quantity of nodules lifted with this system depends on the area swept, the nodule concentration, and the efficiency of the pickup system. Deepsea Ventures plans to lift approximately 140 tons of nodules per hour from waters up to 3 miles in depth. The airlift pump system was tested at 3000-foot depths in 1970 and will be retested at 15,000 foot depths in 1976.

A second hydraulic system is reported to be favored by Kennecott Copper Corporation. This system uses a hydraulic pump to create a continuous upsurge of water in the flexible pipeline extended from the mining ship to the dredgehead on the ocean floor. Nodules are dislodged from the seabed and carried upwards in the suction pipe to the surface, where the minerals are separated from the water. This system was recently tested in 15,000 foot water depths. For their initial operation, Summa Corporation was also believed to be favoring a hydraulic lift system but with a pump located on the ocean floor; however, for later generation mining, they were reported to be planning to use a system which would include submerged barges and underwater tractors [18].

In the third phase of operations—transport—first generation ore carriers are unlikely to do more than transport the nodules, perhaps in crushed and dried form. They are likely to be separate from mining ships, except in the case of the CLB system perhaps, and the nodules will most likely be transferred from hoppers in mining ships to adjacent ore carriers. Second generation transportation vessels might actually complete more processing, so that waste material (e.g., silica) can be separated at the mine site and not transported [19].

In the final steps of mining, there are two basic approaches to the processing of nodules—pyrometallurgical and hydrometallurgical. The pyrometallurgical method involves heating the nodules to break them up, while the hydrometallurgical method involves the use of chemicals to achieve the same purpose. A pyrometallurgical process being developed by Kennecott is the sulfur-dioxide roasting method which converts the metallic oxides in nodules into a more workable sulfate form. The resulting leach liquor of soluble sulfates is treated to achieve a direct precipitation of copper, and using heat and pressure, nickel and cobalt can also be precipitated from the leach. The remaining manganese sulfates are converted to a dry, crystalline manganese sulfate which can be decomposed to yield manganese oxide pure enough for conversion to commercial manganese.

The hydrometallurgical process being developed by Deepsea Ventures involves the use of hydrogen chloride to produce soluble metal chlorides which are leached with water. The leach liquor containing copper, nickel, cobalt, manganese and minor metal chlorides, is separated from the silicate, sulfate and oxides. By a variety of chemical processes, including ion exchange, rehydration and electrolysis, copper, nickel and cobalt are extracted. Manganese in metallic form can also be isolated by additional chemical processes.

In summary, a number of firms in the industrialized nations have invested approximately $200 million and are preparing to invest an additional $800 million to develop the technology and equipment necessary to conduct commercial mining operations at unprecedented water depths. While the earliest date for commercial exploitation of manganese nodule deposits is anticipated to be the end of the decade, negotiators at the Law of the Sea Conference do not have the luxury of waiting for actual mining to begin before attempting to agree upon a regime for ocean mining. The negotiations are now at a critical stage. There is a sense of urgency on the part of the industrialized nations, since potential miners need to be able to guage the investment climate in the 1980s to be able to justify their capital investment; in addition the developed nations are anxious to secure new sources of supply in a world increasingly threatened by cartelization of raw materials [20]. Developing nations, on the other hand, can afford to procrastinate until LDCs, in the aggregate, can negotiate for a share in the benefits to be derived from seabed mining and until developing country producers of seabed minerals, in particular, are assured that their export earnings will not be adversely affected by seabed mining.

LDC PRODUCERS OF SEABED-COMPETING MINERALS

There are two important characteristics of the minerals to be mined from the deep seabed: first, the overwhelming proportion of those minerals will be consumed in the industrialized nations, and second, a very large proportion of the existing supplies of these minerals is produced by developing countries. These two facts are indicative of the underlying disparity in incomes between LDCs and DCs which is at the fulcrum of the controversy between developing and industrialized countries on the future of deep seabed mining. While developing nations tend to consume rather small quantities of copper, nickel, cobalt and manganese (see Table 5–1), they are large producers of copper, cobalt and manganese ore. Land-based developing country producers of these minerals fear that seabed mining will add significantly to supply over time and will therefore, in the absence of a sufficient increase in demand, radically reduce prices. A fall in mineral export prices could produce a severe drop in foreign exchange earnings with accompanying reduction in rates of growth [21]. This poses a potential threat to nations heavily dependent on one or two mineral exports, such as copper in the cases of Zambia, Zaire and Chile (see Tables 5–2 and 5–3).

Copper exports account for approximately 93 percent of Zambia's total exports. This figure has remained fairly constant over time, and there is little likelihood that exports will be diversified in the near future. A reduction in the price of copper, such as the fall in prices from approximately 93¢ per pound in late 1974 to 57¢ per pound throughout 1975, can therefore reduce foreign exchange earnings dramatically. Given the obvious volatility in world copper prices, developing countries heavily dependent on copper exports are concerned that seabed mining or the expectation thereof, will further reduce prices [22].

Chile and Zaire are also heavily dependent on copper exports, as are to lesser degree, Peru, the Philippines and Cyprus. (See Table 5–3). In Zaire, for example, 50 percent of government revenues are provided by the mining sector. Economic models forecast that a fall in the real price of copper results not only in a severe balance of payments deficit and debt repayment crises, but also in significantly

Table 5–1. Metal Consumption *(000 metric tons)*

	1974 Refined Copper[o]	*1974 Nickel*[o]	*1972 Cobalt**	*1972 Manganese Ore**
Developed Countries	6016.9	532.8	22.2	9,850
Developing Countries	488.9	19.6	0.8	2,019
Socialist Economies	1936.5	153.8	4.4	8,834
World Total	8442.3	706.2	27.4	20,703

Sources: [o] *World Metal Statistics* (June, 1975).
*U.S. Bureau of Mines, *Commodity Data Summaries* (Appendix I), 1974, and Bureau of Mines sources.

Table 5–2. World Copper Production *(000 metric tons)*

	1969	*1970*	*1971*	*1972*	*1973*	*1974*
Developing Countries	2,287.0	2,315.3	2,371.6	2,629.1	2,805.3	3,068.5
Zaire	364.0	387.1	405.8	437.3	490.0	544.0
Zambia	719.5	684.1	651.4	717.7	706.6	698.0
Philippines	131.4	160.3	197.4	213.7	220.3	225.3
Chile	688.1	691.6	708.3	716.8	735.4	902.1
Peru	199.0	212.1	212.9	217.0	220.0	213.2
Others	185.0	180.1	195.8	326.6	433.0	626.4
Developed Countries	2,550.8	2,850.5	2,776.7	3,015.0	3,249.1	3,171.9
U.S.	1,401.2	1,560.0	1,380.9	1,510.3	815.1	1,448.8
Canada	520.0	610.3	654.5	719.7	1,566.7	826.2
South Africa*	152.6	167.0	174.3	183.4	203.1	205.2
Australia	131.1	157.8	177.3	180.5	215.1	255.5
Japan	120.3	119.5	121.0	112.1	91.3	82.1
Others	225.6	235.9	268.7	309.0	357.8	354.1
Free Market Total	4,837.8	5,165.8	5,148.3	5,644.1	6,054.4	6,240.4
Socialist Economies Total	1,111.5	1,204.1	1,315.3	1,416.0	1,475.0	1,668.7
Total World	5,949.3	6,369.9	6,463.3	7,060.1	7,529.4	7,909.1

*Includes Southwest Africa.

Source: World mine production figures, *World Metal Statistics*.

Table 5–3. Developing Countries—Copper Exports *(mil US $)*

	1969			1970			1971		
	Copper Exports	*Total Exports*	$\frac{CE}{TE}$ (%)	*CE*	*TE*	$\frac{CE}{TE}$	*CE*	*TE*	$\frac{CE}{TE}$
Zaire	429	649	66	494	742	67	415	695	60
Zambia	1,011	1,073	94	954	1,001	95	630	679	93
Chile	973	1,020	95	910	1,178	91	701	962	73
Peru	259	863	30	269	1,044	26	170	891	19
Philippines	133	855	16	185	1,062	17	185	1,128	16
Cyprus	18	98	18	22	109	20	15	123	12

Table 5–3. continued

	1972			1973			1974		
	CE	*TE*	$\frac{CE}{TE}$	*CE*	*TE*	$\frac{CE}{TE}$	*CE*	*TE*	$\frac{CE}{TE}$
Zaire	406	691	59	618	1,004	62	953	1,303	73
Zambia	687	758	91	1,090	1,158	94	1,303	1,406	93
Chile	631	855	74	1,026	1,231	83	1,654	2,481	67
Peru	188	943	20	284	1,039	27	348	1,511	23
Philippines	200	1,105	18	281	1,788	16	376	2,636	14
Cyprus	14	134	10	18	175	11	16	152	11

Sources: IMF, *International Financial Statistics.*

reduced rates of consumption, savings, and income growth [23]. A severe fall in the price of copper in 1975 has been a prime cause of the economic difficulties of Zaire, Zambia, Chile and Peru, nations whose economies are so shaken that international debt repayments will need to be postponed [24]. Given the serious consequences for LDC copper producers of price reductions, for example, any examination of deep seabed mining should include an analysis of the potential adverse effects of seabed mining on the economies of land-based mineral producers. Such an examination is contained in Chapter 6.

While the role of developing country producers is pre-eminent in copper production, LDC land-based sources are also important in the supply of manganese ore. Table 5–4 reveals the extent of the role of LDCs in manganese ore production. In 1973, LDCs produced 6.77 million metric tons of manganese ore, which is 54 percent of the market economy total and 31 percent of the world total. The largest LDC producers of manganese ore are Gabon, Brazil and India; however, only in the case of Gabon is manganese a major source of foreign exchange earnings (see Table 5–5). It should also be noted that the relative importance of manganese in Gabon's export profile has been markedly reduced in the past few years (viz., manganese ore exports were 27.5 percent of total exports in 1972 but accounted for 11.7 percent of total exports in 1973), because of the level of Gabonese petroleum exports. In comparison with Gabon's current manganese exports earnings (estimated to be close to their 1973 level of $42 million), total revenue derived from the export of oil was $804 million in 1974 [25]. It can be argued, therefore, that manganese ore production does not critically contribute to the stability and growth of any particular LDC economy to the extent that we can predict at the present. It should be noted, however, that we are ignoring the possible effects of future seabed mining on the decision to begin exploitation of existing land-based manganese ore deposits and are concerned with current LDC producers [26].

Cobalt is a by-product of copper and, at times, nickel mining. The cobalt market is extremely small; total output in 1974 amounted to a mere 29.9 thousand metric tons. The cobalt market is thus approximately four one-hundredths the size of the world copper market. The case of cobalt is perhaps the most striking in terms of the non-complementarity of interests of the consumers of cobalt—almost exclusively developed nations—and of the land-based producers—primarily developing countries and in particular Zaire. Developing countries consume 3 percent of total world cobalt production and

Table 5–4. World Production of Manganese Ore *(000 metric tons)*

	1970	*1971*	*1972*	*1973*	*1974*[p]
Developing Countries	6,287.3	7,703.6	6,945.6	6,771.4	5,873.2
Mexico	269.6	262.7	291.0	358.3	n/a
Brazil	1,849.4	2,561.1	2,025.3	2,123.6	1,786.0
Gabon	1,430.6	1,837.3	1,906.4	1,888.8	2,104.8
Ghana	399.0	589.2	490.5	313.2	n/a
Morocco	110.6	99.9	94.6	143.9	n/a
Zaire	341.5	380.9	363.7	328.7	n/a
India	1,625.2	1,809.9	1,616.3	1,511.0	1,424.3
Other	261.4	162.6	157.8	103.9	558.1
Developed Countries	3,720.7	4,580.3	4,675.9	5,839.0	6,718.1
South Africa	2,637.6	3,185.9	3,220.3	4,110.3	3,687.2
Japan	266.2	280.5	256.7	185.8	n/a
Austria	739.0	1,033.8	1,149.7	1,498.6	1,498.5
Other	77.9	80.1	49.2	44.3	1,532.4
Socialist Economies	7,929.2	8,406.6	8,908.2	9,080.9	9,555.1
Total World	17,937.3	20,690.5	20,529.7	21,691.3	22,146.4

n/a = not available.
[p]preliminary.
Sources: Bureau of Mines, *Minerals Yearbook* (1973, 1974) and *Commodity Data Summaries*, Appendix I (1974, 1976).

Table 5–5. Developing Countries–Manganese Ore Exports *(mil US $)*

	1970			1971		
	Manganese Ore Exports (ME)	*Total Exports (TE)*	*ME/TE (%)*	*ME*	*TE*	*ME/TE*
Brazil	31	2,191	1.4	38	2,728	1.4
Gabon	12	121	9.9	36	198	18.2
India	19	2,006	0.9	14	2,124	0.7
Ghana	7	451	1.6	6	598	1.0
Zaire	6	788	0.8	4	695	0.6
Morocco	5	490	1.0	4	538	0.7

Table 5–5. continued

	1972			1973			1974		
	ME	*TE*	*ME/TE*	*ME*	*TE*	*ME/TE*	*ME*	*TE*	*ME/TE*
Brazil	27	3,795	0.7	19	6,199	0.3	50	7,260	0.7
Gabon	53	193	27.5	44	360	11.7	n/a	1,058	n/a
India	11	2,501	0.4	11	2,753	0.4	11[1]	3,933	0.3
Ghana	8	659	1.2	8	843	0.9	n/a	1,001	n/a
Zaire	5	689	0.7	2	840	0.2	5	1,303	0.4
Morocco	6	619	1.0	8	868	0.9	11	1,782	0.6

n/a = not available. [1]Fiscal year April—March.

Sources: Agency for International Development.
IMF, International Financial Statistics, 1974.
Gabon, *Bulletin Mensual du Statistique*, 1974.

account for 76 percent of world production themselves. Of the three-quarters of world cobalt produced by LDCs, Zaire alone produced over 17,000 metric tons or 70 percent of that amount (see Table 5–6). Cobalt exports were 8.7 percent of total Zairian exports in 1974 (see Table 5–7).

A large increase in cobalt supply would probably reduce prices substantially, since world cobalt demand is rather small and generally thought to be rather price inelastic [27]. Zaire, a price setter for cobalt, would be the country most seriously affected by such additions to world cobalt supply.

Another mineral to be exploited from the deep seabed in quantities sufficient to concern land-based producers is nickel. Nickel production is concentrated in three countries—Canada, New Caledonia and the Soviet Union. Together, these countries account for approximately two-thrids of world production. Total developing country production of nickel was 52,000 metric tons in 1974, or 7 percent of the world total (see Table 5–8). Developing countries are also rather small consumers of nickel, accounting for less than 3 percent of world consumption in 1974. Recently, however, nickel production has been started in Indonesia and the Dominican Republic. Indonesia's output of 16,000 tons in 1974 is less than 2 percent of the world total, and although export value data are unavailable, it can be estimated to be perhaps 1.5 percent of total export earnings. As a member of OPEC and a major oil exporter, Indonesia's dependence on nickel exports is miniscule and should remain so through 1985. The Dominican Republic, where nickel production began in 1972, produced 31,200 tons of nickel in 1974, worth $93 million, which accounted for 14.6 percent of total export earnings. Dominican nickel production, approximately 4 percent of the world total, is therefore an important new source of foreign exchange.

In summary, it is likely that a handful of developing nations have legitimate concerns about the effects of seabed mining on the development of their domestic economies. In particular, producers of copper such as Zambia, Zaire and Chile rely on copper exports as the major source of foreign exchange. Zaire also derives a significant portion of its total export earnings from cobalt production. Whereas Gabon might have been added to this list of possible "affected countries," the discovery of onshore and offshore oil deposits coupled with the quadrupling of oil prices in recent years greatly increases the earnings capacity of that country. Therefore a change in manganese ore export earnings is unlikely to exert a major impact on Gabon's economy.

The uncertainty about the future effects of seabed mining on LDC

Table 5–6. World Cobalt Production *(metric tons)*

	1970	*1971*	*1972*	*1973*	*1974*[p]
Developing Countries	16,960	17,331	17,820	19,381	22,780
Zaire	13,960	14,289	12,906	14,846	17,271
Zambia	2,400	2,080	3,292	3,090	3,112
Morocco	600	962	1,577	1,399	1,725
Other	—	—	45	46	672
Developed Countries	4,080	3,967	3,501	3,495	3,835
Canada	2,070	1,930	1,497	1,493	1,893
Australia	470	783	758	757	759
Other	1,540	1,254	1,246	1,245	1,183
Socialist Economies	3,573	3,080	3,126	3,260	3,304
Total World	23,850	24,378	24,447	26,136	29,919

[p]preliminary.

Sources: U.S. Bureau of Mines, *Commodity Data Summaries* (Appendix I), 1974, 1976.

Table 5–7. Developing Countries—Cobalt Exports *($ US mil)*

	1971			1972			1973			1974		
	Cobalt Exports (Co E)	*Total Exports (TE)*	*Co E / TE (%)*	*Co E*	*TE*	*Co E / TE*	*Co E*	*TE*	*Co E / TE*	*Co E*	*TE*	*Co E / TE*
Zaire	46	695	6.6	53	689	7.7	108	1037	10.4	114	1303	8.7
Zambia	6	679	0.9	12	758	1.6	7	1158	0.6	12	1406	0.8
Morocco	1	538	0.2	4	619	0.7	4	845	0.5	11	1782	0.6

Sources: Agency for International Development.
IMF, International Financial Statistics.
Statistiques du Commerce Exterieur, Zaire, 1969–1972.

Table 5–8. World Nickel Production *(000 metric tons)*

	1969	*1970*	*1971*	*1972*	*1973*	*1974*
Developing Countries	6.6	13.9	16.8	13.8	49.8	52.2
Dominican Republic	—	—	.2	17.4	30.1	31.2
Indonesia	4.9	10.8	12.0	14.1	15.8	16.0
Other	1.7	3.1	4.6	4.3	3.9	5.0
Developed Countries	359.7	496.5	503.2	430.9	443.2	526.1
Canada	193.8	277.5	267.0	234.9	241.2	262.2
New Caledonia	117.0	138.5	150.9	105.0	108.1	136.0
Australia	11.2	29.8	31.1	35.8	35.7	54.5
Other	37.7	50.7	54.2	55.2	58.2	73.4
Socialist Economies	149.0	157.0	163.4	159.7	160.0	160.7
U.S.S.R.	105.0	110.0	120.0	120.0	115.0	120.0
Cuba	37.0	40.0	35.0	32.0	32.0	36.5
Other	7.0	7.0	8.0	7.0	12.0	4.2
Total World	515.3	667.4	683.4	626.4	653.0	739.0

Source: *World Metal Statistics.*

land-based producers has been one of the major obstacles in the Law of the Sea negotiations on the deep seabed. Influential delegations such as Zaire and Chile have effectively rallied the support of other developing nations to press for international control over the pace of seabed mining. Inasmuch as LDCs are not major consumers of seabed affected minerals, they do not foresee themselves gaining any significant benefits from seabed mining by the industrially advanced countries and are therefore wont to internationally sanction, in a Seabed Treaty for example, the rights of the industrialized states to mine the deep seabed at will [28]. One of the crucial issues of examination, therefore, is the extent to which LDC land-based producers will actually be affected and whether or not measures can be instituted to compensate their losses, should they be substantial. Thus, our next task is to develop price forecasts for seabed minerals, within the context of an examination of the economic viability of seabed mining investments. In Chapter 6 we analyze the potential revenues, costs and profits of manganese nodule mining and the effects of mining on land producers of minerals.

NOTES TO CHAPTER 5

1. See pp. 181–182 for discussion.

2. This section is based primarily on John L. Mero, *The Mineral Resources of the Sea*, Amsterdam: Elsevier, (1965). We also benefited from discussions with Arnold J. Rothstein of Deepsea Ventures, Inc.

3. While mining in the continental margin (excluding hydrocarbons) is small in magnitude (i.e., its 1970 value was estimated at $180 million) a number of minerals including tin, barites, sulphur, iron sands, diamonds, calcium carbonate and sand gravel are commercially exploited. See "Deep Ocean and Offshore Mining", *Mining Engineering* (April 1975).

4. A prime site can be described as being topographically smooth, with a dense concentration of nodules with small variations in size, content, and specific gravity. The mineral content of the nodules should clearly be high in specific minerals whose extraction is likely to yield high revenue. In the current situation, miners are primarily interested in deposits with high nickel and copper content.

5. Mero, *op. cit.*, pp. 154, 175–177.

6. *Ibid.* pp. 173–175.

7. *Ibid.* pp. 175.

8. See note 4. It should be borne in mind that an average sized mine site in the case of seabed mining might be 30,000 square kilometers. Such an area can contain sufficient deposits to sustain full-scale mining for at least forty years. See "Deep Ocean and Offshore Mining."

9. Mero, *op. cit.*, pp. 225–230. Estimates are based on a detrital-mineral-free basis. Detritus are the fragments of rock produced by the disintegration or wearing away of the rock. Detrital-mineral free (dmf) excludes any such accumulations from the nodule assay. It should be stressed that dmf is a geological measure which often underestimates the economic value of assayed minerals, since detritus can be commercially used in some mineral processing.

10. D. Horn, B. Horn and M. Delach, "Ocean Manganese Nodules: Metal Values and Mining Sites," unpublished manuscript, Technical Report No. 4, National Science Foundation, 1973, Washington, D.C.

11. See Deepsea Ventures, "Notice of Discovery and Claim of Exclusive Mining Rights," and "Request for Diplomatic Protection and Protection of Investment," (November 15, 1974).

12. The claim was rejected. The US appears still to be committed to a multilateral negotiated agreement on deep seabed mining jurisdiction; however, there are some indications that if agreement on an international framework for seabed mining rights is not forthcoming, the United States might begin to license its miners nationally to exploit resources in the deep seabed.

13. Deepsea Ventures, *op. cit.*

14. This section draws heavily on the following sources: (1) "U.S. Draft Environmental Impact Statement," UN Third Conference on the Law of the Sea (1974); (2) M.A. Dubs, J.E. Flipse and R.J. Greenwald, "Pre-Production Manganese Nodule Mining Activities and Requirements," in *Mineral Resources in the Deep Seabed* (1973); (3) J.L. Mero, *op. cit.*; (4) N. Cornell, "Techno-Economic Focus Paper on Manganese Nodule Mining," Brookings Institution (April, 1973); and (5) UN Third Conference on the Law of the Sea, "Economic Implications of Seabed Mineral Development in the International Area," A/Conf. 62/25 (May 22, 1974).

15. Nations directly involved in nodule mining ventures include Belgium, West Germany, France, Canada, Great Britain, Japan, and the United States. The USSR is also reported to be exploring deep seabed mining possibilities.

16. See note 4.

17. See UN, A/Conf. 62/25 (May 22, 1974) for details.

18. Although Summa is no longer reported to be in the nodule mining field, some of these ideas for second generation mining technologies are possibilities. See "Deep Ocean and Offshore Mining," for a good summary of deep seabed mining techniques.

19. Such methods involve as yet undetermined environmental costs.

20. See C. Fred Bergsten, "The New Era in World Commodity Markets," *Challenge*, vol. 17, no. 4 (Sept.–Oct., 1974) pp. 34–42.

21. See, for example, C. Reynolds, "Domestic Consequences of Export Instability," *American Economic Review*, Vol. 53, No. 3 (July 1965); B. Massel, "Export Instability and Economic Structure," *American Economic Review*, Vol. 58, No. 4 (September 1970); S. Naya, "Fluctuations in Export Earnings and Economic Patterns of Asian Countries," and C. Glezakos "Export Instability and Economic Growth, A Statistical Verification," *Economic Development and Cultural Change*, Vol. 21, No. 4 (July 1973); and P. Kenen and C. Voivodas, "Export Instability and Economic Growth," *Kyklos*, Vol. 25, No. 4, (1972).

22. It should be noted that the doubling of copper prices between 1972 and 1974 was a result, among other reasons, not only of boom conditions in copper-using industries, but also of speculation and a general flight from currency into commodities. The recession in 1975 and the increase in copper production in response to high prices was in large measure responsible for the dramatic fall in prices.

23. Multilateral Institution sources.

24. See A. Crittenden, "U.S. Bank Loans Abroad Stirring Concern," *The New York Times* (January 15, 1976).

25. International Monetary Fund.

26. It is a fact that countries contemplating investments in manganese ore production have been wary about the effect of seabed mining on the price of manganese in the 1980s and thus about the advisability of the investment. Also the degree of dependence of an economy on manganese ore exports can obviously be quite different a decade from now.

27. See Chapter 6, notes 21 and 61.

28. Essentially the LDCs wish to control seabed mining output, not only to protect LDC land-based producers, but also to appropriate much of the economic rent involved in mining production. See pp. 188–189.

Chapter 6

The Economics of Manganese Nodule Exploitation

INTRODUCTION

The viability of a mining investment—once the characteristics of the resource deposit are known and the technology is available—depends on the calculation of expected costs and returns. What renders such an investment decision difficult and unique in the case of manganese nodules is the fact that it is necessary to forecast both future minerals prices and future costs for 1985, an undertaking which is rather speculative at this juncture. In addition, the investment climate (i.e., the nature of the international deep seabed regime) has not yet been established, and thus mining consortia have invested approximately $200 million to date under conditions of extreme uncertainty concerning the feasibility of exploiting resources which lie predominantly in international areas [1]. At the same time, the international community, including the land-based mineral producers, is in the process of negotiating a Seabed Treaty, which would determine for decades to come the rules and regulations of deep seabed mining. It is remarkable, yet unavoidable, that these negotiations must proceed in the absence of any experience with manganese nodule mining and in almost complete ignorance of the full economic implications of nodule mining.

At the risk of muddying the waters, this chapter attempts to produce some reasonable minerals price forecasts for the seabed mining era and thus to estimate the potential revenue generated from mining. The second section discusses the cost of manganese nodule mining. In section three, we attempt a cost-return analysis in 1974 constant dollars and report profit estimates. The penultimate section examines the possible adverse impact of manganese nodule mining on LDC mineral producers. Last, we review existing compensatory financing facilities which might be used to assist adjustment in LDCs whose economies are affected, and we analyze the possible uses of international seabed taxation, should an international tax be imposed on nodule miners [2].

REVENUE PROJECTIONS

Revenue projections depend on the assumed future levels of production of deep seabed mining and on the forecasts of future metals prices. Estimates of this kind are by their nature speculative; they depend intrinsically on the underlying assumptions concerning future production from land-based and seabed sources of supply and on future world demand for particular minerals and their substitutes. Supply uncertainties center on the number of seabed mining operations, the number of new land sources of supply and the production levels of traditional producers. Demand uncertainties concern the level of demand at pre-seabed production prices, which is dependent on world output, prices of substitute metals, and technological considerations and tastes, and the price responsiveness of demand when seabed production augments supplies of minerals such as copper, nickel, cobalt and manganese [3]. In a milieu in which copper prices can double and fall back to their original level in the space of a few years (1972–75) and in which new substitutes and technologies abound, any examination of minerals prices a decade ahead must be viewed as illustrative.

The first questions to be addressed are how many mining sites will be operational by, say, 1985 and what will be the level of production at each mine site. An examination of the industry and discussions with others attempting to forecast the parameters of manganese nodule mining lead us to suggest that the volume of mining in 1985 may range between 13 million and 17 million dry tons per annum [4]. It is assumed that Deepsea Ventures will initiate mining at a level of production of one million tons of nodules per annum by 1980, and we assume this output is maintained until 1985. Kennecott-led and INCO-led consortia plan three million ton per annum operations to begin in the early part of the next decade but certainly before 1985. Later ventures involving other American, Japanese, European and Soviet firms or entities are predictable, all within a one to three million ton per year scale of operations. For our analysis, let us assume that nodule mining will produce 15 million dry tons of output in 1985, a figure corresponding to a recent United Nations estimate [5].

It is reasonable to expect that initial mining will occur in prime sites in the Pacific, areas in which nodules contain an average of 25 percent manganese, 1.5 percent nickel, 1.2 percent copper, .25 percent cobalt and small quantities of trace metals such as vanadium, molybdenum and zinc. Only Deepsea Ventures has indicated a desire at this point to extract the trace metals from nodules [6]. Trace metals might be recovered in quantities of perhaps .11 percent vanadium, .08 percent zinc and .15 percent molybdenum per nodule mass [7]. An average rate of recovery of minerals can be assumed to be 95 percent [8]. We assume for our analysis that of the 15 million tons of nodules mined in 1985, only the one million tons exploited by Deepsea Ventures will involve the extraction of manganese—and in that case most probably in its metallic rather than ore form—and the trace metals. The decision to extract seven metals is unique to Deepsea at this point, and other firms, for example Kennecott, have indicated that they are primarily interested in extracting the three primary minerals—nickel, copper and cobalt [9].

Column 2 of Table 6–1 reports the approximate amounts of seven metals contained in a ton of dry nodules. Column 3 reports the net metal yield for 15 million tons recovery under our assumptions. Column 5 compares these estimated production figures for 1985 to current (1974) world production. For the primary three metals, these calculations are adjusted to compare nodule-source production to estimated 1985 consumption demand. It appears that the increase in copper supplies in 1985 as a result of manganese nodule mining will not be significant. Nodule-source production of nickel, and more particularly cobalt, will be substantial relative to total production, however. The cobalt market would indeed be glutted if land-based producers maintain their output levels when nodule-source cobalt augments total supply [10]. Cobalt supplies might be increased by two-thirds with the advent of nodule mining, and nickel supplies might be increased by almost one-sixth. If manganese metal is extracted as planned, seabed manganese output would more than double current supply and significantly add to 1985 supply, with obvious consequences for the price of manganese metal [11]. With respect to trace metals, it appears that seabed mining will not exert a significant impact on metal markets; this is predicated on the assumption that only one firm has announced its intention to extract trace metals. If, however, given the joint production nature of nodule

Table 6–1. Production Profile

Metals	*(1) Metal Content per Dry Ton of Nodules*	*(2) Pounds of Metal Output per Dry Metric Ton of Nodules*	*(3) Metal Produced from 15m. Metric Tons of Nodules (000 Metric Tons)*	*(4) 1974 World Production (000 Metric Tons)*	*(5) Seabed Source as a Proportion of Total Current (1985 Est.) Production*
Nickel	1.5%	33.6	225.0	707.1[a]	31.8% (14%)
Copper	1.2%	26.9	180.0	7909.1[b]	2.3% (1.9%)
Cobalt	.25%	5.6	37.5	29.9[c]	125.4% (62.5%)
			*Metal Produced from 1m. Tons of Nodules (000 Metric Tons)**		
Manganese (metal)	25.0%	560.0	250.0	223.4[d]	111.9%
Vanadium	.11%	2.5	1.1	20.1[e]	5.5%
Zinc	.08%	1.8	.8	5700.0[e]	.01%
Molybdendum	.15%	3.4	1.5	83.6[e]	1.8%

*Assume that only one operation of one million tons a year extracts manganese and trace metals.

Sources: [a] *World Metal Statistics*, June 1975.
[b] *World Metal Statistics*, June 1975.
[c] U.S. Bureau of Mines, *Commodity Data Summaries*, 1976 and UN A/Conf. 62/25 (May 74).
[d] 1972 figure calculated from UNCTAD, TD/B/C.1, Cons. F/L2 (Feb. 1974) and Cornell *op. cit.*
[e] U.S. Bureau of Mines, *Commodity Data Summaries*, 1976.

mining, other firms extract these particular metals, supplies might be significantly affected, especially for vanadium, possibly resulting in price reductions. A more detailed analysis of metals prices follows.

Copper is probably the most widely used mineral of those to be mined from the deep seabed. Copper is an excellent conductor of electricity, with high corrosion resistance and many alloying possibilities, and demand is likely to grow steadily in the next ten years. The World Bank projects a 3 to 4 percent annual growth rate in copper demand until 1980, and an approximate growth rate of 4 percent annually thereafter until 1985 [12]. These estimates of demand growth are based in part on the finding that the elasticity of copper demand with respect to industrial production of the developed countries is on average between .80 and 1.0 [13]. With 1974 world copper production totalling approximately 7.9 million metric tons and known world reserves reported to be 370 million tons (with an identical amount thought to exist in undiscovered land reserves), the outlook for steadily expanding copper production is quite optimistic [14]. The World Bank estimates that land-based mine capacity will rise by approximately 6 to 9 percent annually between 1972 and 1980 in the absence of seabed mining, and thus suggests that real copper prices will be reduced somewhat over time [15]. While the dramatic (i.e., 50 percent) decline in copper prices between 1974 and 1975 makes forecasting obviously precarious, experts tend to place the 1985 copper price at approximately 75¢/lb. in 1974 prices [16]. With demand projected to be 10.3 million tons in 1985, the proportion of total world production of copper to be supplied by seabed sources is approximately 1.9 percent. Therefore, in selecting an estimated copper price for revenue forecasting, we assume for our analysis that the market price of copper is unaffected by seabed production.

Cobalt, a by-product of copper and nickel production, is a highly heat-resistant metal which possesses chemical and magnetic properties which are conducive to expanded industrial uses. Cobalt consumption is increasing worldwide, particularly in the United States where cobalt consumption increased by 18.5 percent from 1972 to 1973, primarily because of its introduction into the manufacture of automobile anti-pollution devices. The largest international supplier of cobalt is Zaire, which in 1974 mined 58 percent of the world's cobalt. The cobalt market itself is rather small, with a total production in 1974 of less than 30,000 tons. Even with forecasted annual increases in demand of approximately 6 to 7 percent, consumption is not likely to exceed 60,000 tons by 1985 [17]. The amount of cobalt recoverable from the mining of 15 million tons of manganese nodules of the assays previously assumed totals 37,500 tons, or 62.5 percent of projected 1985 demand. It is probable, therefore, that cobalt prices will fall as a result of nodule mining unless Zaire is willing to reduce substantially its cobalt production in the 1980s. In either case one should expect Zairian cobalt export revenues to fall, either because of reduced world prices or through voluntary output restraint or a combination thereof. Cobalt and nickel are close substitutes in many industrial uses, so that if cobalt prices fell to the level of nickel, its closest available substitute, sufficient demand should exist to stabilize cobalt prices at a level fairly close to the price of nickel. Since the cobalt market is extremely small when compared to the volume of demand for nickel, it is assumed that cobalt prices will stabilize at the price of nickel in 1985.

Nickel is the key metal which has stimulated the interest in deep seabed mining, and nickel revenue is thus of paramount importance to the viability of a seabed mining venture. On the supply side, nickel production is concentrated in three countries—Canada, New Caledonia and the Soviet Union. Canadian nickel production, controlled by the International Nickel Company of Canada, Ltd. (INCO), accounted for almost all market economy nickel output in 1950. This dominant position is being gradually eroded by a second major producer, Société Le Nickel of New Caledonia (viz., INCO accounted for 73 percent of total production in 1950 and 58 percent of total world production in 1970). In 1973, INCO and Le Nickel controlled more than half of world production of nickel and two-thirds of the production of nickel in market economies. Developing-country nickel producers include the Dominican Republic with 4 percent of the world nickel total and Indonesia producing close to 2 percent of the world total. Cuba produced 5 percent of total world nickel production in 1974.

Nickel ores are found in the form of either laterites or sulfides [18]. Laterites are soft-surface deposits, the mining of which requires a technology of the earth-moving type. Laterites tend to be found in tropical and sub-tropical climates and are mined in New Caledonia, for example. Sulfides, nickel ores containing sulfur, are hard rock deposits, where mining requires deep drilling and tunneling. Canadian deposits are of the sulfide form. Inasmuch as Canadian mines are close to sources of hydroelectric power, whereas Caledonian mining requires more expensive energy sources, Canadian nickel is cheaper to produce. While it is not expected that nickel produced from seabed mining will be lower in cost than that produced by INCO, firms investing in seabed mining hope to produce nickel at costs lower than New Caledonian sources. Since land-based nickel reserves are abundant, producing lower cost nickel is important for the viability of seabed mining. In other words, if nodule-source nickel can be produced at a cost lower than Le Nickel's, then seabed produced nickel would successfully displace (or partially displace) Le Nickel's output [19]. Seabed miners have an added advantage, of course; they are producing nickel as a joint product with at least copper and cobalt and can thus spread some of the cost of production among various minerals [20].

Before we discuss possible future nickel prices, a review of the supply and demand conditions of the nickel market may prove instructive. Nickel is produced in an oligopolistic market (i.e., a market in which there are a few sellers and in which the individual seller's output affects the market price), in particular, in a market in which the major producers, INCO and Le Nickel, maintain such a large share of supply that they can determine market prices. To be an effective "price-setter," a firm (or group of firms) should face a highly inelastic demand for its output. In the case of nickel, the price inelasticity of demand is based on the fact that nickel is extremely difficult to substitute for, at least in the short to medium term and thus the quantity of nickel demanded is not likely to change greatly with a change in nickel prices [21]. To forecast the effect of new entrants (i.e., nodule-source production) on the nickel market, we need to know the relative production capacities of producers, their cost structure, and their pricing policies [22]. If we assume that seabed nickel production will account for approximately 14 percent of total 1985 nickel demand (an assumption to be substantiated later), and that the cost of mining nickel from manganese nodules will be less than the $1.92 per pound reported cost of producing lateritic ores, seabed producers stand to be able to compete favorably with Le Nickel and other laterite producers for a share of the nickel market [23]. If we assume collusive behavior on the part of the dominant producers, Le Nickel, INCO or others could reduce their respective nickel outputs to accommodate the seabed entrants and maintain a desired level of supply. Should land-based producers not wish to reduce production, nickel would be supplied in amounts exceeding normal (i.e., pre-seabed mining) levels, and given a static demand structure, prices would fall. They would tend to fall faster, the more price inelastic the demand for nickel [24]. Given the method in which nickel prices are currently set and the fact that INCO is one of the three firms in the nodule mining field, it seems that nickel prices will not be radically affected by seabed mining and that the new entrants will be accommodated into the nickel-producing group.

Therefore, it is reasonable to expect that the production of nickel from manganese nodules *per se* will not lead to a major real reduction in 1985 nickel prices.

If it is essentially correct that nodule-source nickel production will not disrupt the nickel market, then it is useful to proceed to determine what the price of nickel might be in context which ignores deep seabed mining. On the demand side nickel has no major direct substitutes; nickel is crucial in the production of stainless steel (viz., 40 percent of all nickel produced is for stainless steel production), but it is also a small item in the cost structure of stainless steel producers. This, plus the fact that stainless steel is to some extent also produced by a limited number of firms which tend to be better able to pass cost increases on to consumers; is consistent with the view that the demand curve for nickel is price inelastic. A recent econometric model of nickel does show that nickel consumption in the United States rises with increases in the prices of cobalt and aluminum, the latter with a lag of two to three years, however [25]. Sources also report an estimated elasticity of nickel demand with respect to income of unity (1.0) and with respect to industrial production of 1.5–2.7. The World Bank estimates that nickel demand will increase by a 6.2 percent per annum between 1975 and 1980 [26]. If we accept a 6 percent across-the-board growth rate in demand until 1985, world consumption would total 1.59 million tons in 1985. Given the income elasticity of nickel demand, this consumption figure is attainable with a 6 percent average real growth rate of world output. At a 15 million ton level of nodule production, the seabed-source output of nickel would then account for 14 percent of total production.

On the supply side, we know that nickel reserves are abundant and that additional nickel projects are being planned or considered in many countries [27]. The major current constraint to nickel production appears to be the high cost of mining, especially the cost of energy, in particular for drilling at deeper levels.

In order to empirically estimate demand and supply functions for nickel and forecast prices, we construct a simple two equation simultaneous econometric model, described in detail below [28]. To summarize the findings briefly, the hypothesis that the demand curve for nickel is perfectly price inelastic is empirically tested and accepted. It is also determined that the demand for nickel depends primarily on the level of steel output. On the supply side, it is found that price does positively influence nickel production and that rising costs of mining machinery and equipment significantly reduces nickel production, other factors being held equal.

Based on this model, we can derive a 1985 price in 1974 dollars under assumptions of inelastic demand with respect to price if hypothesized parameters are introduced for the following: (a) growth in total world steel production annually; (b) movement of the price index of mining machinery and equipment annually; and (c) the rate of discount to evaluate the present value of a future nickel price [29]. Table 6–2 reports estimated nickel prices given 4 percent and 6 percent annual growth rate assumptions for steel output and mining machinery prices and with the assumption that the rate of machinery price inflation corresponds to the average rate of world price inflation. For example, assuming a 4 percent average annual worldwide rate of inflation of input prices, a 4 percent rate of discount, and a 4 percent average annual rate of growth in steel production yields a 1985 nickel price forecast of $2.83/lb. in 1974 dollars in the absence of seabed mining. A growth rate in steel production of 6 percent per year between 1973 and 1985, when coupled with a 4 percent annual increase in mining equipment prices yields a 1985 nickel price (discounted by 4 percent per year) equal to $3.22 in 1974 units of account [30]. The projected level of nickel consumption in 1985 under these assumptions is 1.45 million tons, which is equal to an average 5.9 percent increase in nickel consumption per year between 1973 and 1985 [31].

It is proposed to subject the model to a change in supply in 1985, namely, an addition of 225,000 metric tons of nickel to the forecast land-based supply of 1.45 million tons (achieved with a 5.9 percent growth rate in consumption demand over time). Given the estimated supply elasticity (evaluated at 1985 prices and quantities) of .40, and an infinitely price inelastic demand, we calculate the percentage reduction in prices per one percent addition to supply to be .39 [32]. Since seabed-produced nickel would add 16 percent to the normal 1985 nickel supply, we would expect prices to fall by 6.25 percent, from $3.22 per pound in the absence of seabed production to $3.02 per pound with seabed production. This price estimate is considered to be a maximum price, since the model is based upon

Table 6–2. 1985 Real Nickel Price Forecasts (1974 $/lb.)

		Growth Rate of World Steel Output (Average per Annum)	
		4%	6%
Growth Rate of an Index of Mining Equipment Prices (Average per Annum)	4%	$2.83	$3.36
	6%	$3.22	$3.65

totally price inelastic demand, an assumption which tends, *ceteris paribus*, to yield high base price estimates [33].

An alternative and more sophisticated econometric model, constructed by Professor F.G. Adams for UNCTAD, yields a 1980 nickel price estimate in current dollars of $2.38/lb. under similar inflation and growth assumptions [34]. Unfortunately, all of the projections commissioned by UNCTAD suffer from the assumptions that seabed output will begin in 1974 and reach a peak in 1980. The important period of examination, 1980–1985, is not covered. Adams' estimated 1980 nickel price in the absence of nodule mining is $2.43/lb. and this falls to $2.40/lb. with the addition of one mining venture (viz., production of one million ton nodules) in 1980 [35]. Adams' estimates the elasticity of nickel prices with respect to additional nickel output from seabed sources to be between .7 and 1.0 [36]. The lower response occurs with as much as ten million tons of nodule mining per year at the end of the period. If we make the basic assumption that in the absence of seabed output, the real price of nickel would remain stable at $2.43/lb. until 1985, the model can then be subjected to the same exogenous shock of an added 15 million tons of nodules (viz., this corresponds to an average of Adams' two highest production assumptions). An increase of 6 percent per year in nickel consumption between Adams' 1980 base year and 1985 yields a total level of nickel output of 1.23 million tons in 1985; seabed output of nickel of 225,000 tons would thus add 18 percent to total supply and might reduce prices by 12.6 percent or a level of $2.12/lb. in 1974 prices.

Given the crucial role of the price of nickel in the calculation of revenue to be derived from seabed mining, and the uncertainty surrounding any particular price estimate, it is prudent to specify two alternatives, a low 1985 price of $2.00/lb. and a high 1985 price of $3.00/lb. This seems advisable since the nature and stability of the demand for nickel vis-à-vis price on the consumer side and the marketing arrangements and degree of competition in nickel supply on the producer side are variable within the space of a decade. In forecasting revenue; the 1985 price of cobalt is assumed to be equal to the 1985 price forecast for nickel.

Last, we turn to the case of manganese, the most abundant metal in manganese nodules. Were all miners operating in 1985 to extract manganese ore, production would increase by some 3.75 million metric tons of ore, or approximately 10 percent of projected 1985 production [37]. Only one deep seabed mining firm, Deepsea Ventures, has announced its intention to extract manganese from the nodules, however, and Deepsea anticipates extracting manganese in its metallic form. Recent data indicate that manganese metal consumption was approximately a quarter of a million metric tons [38]. Despite growth in consumption demand over time, an additional quarter of a million tons of manganese metal would be likely to glut the market for manganese metal. It is not likely that the present (mid-1974) price for electrolytic manganese (the end use of manganese metal) of $.42/lb. could be maintained in real terms. Inasmuch as electrolytic manganese can be substituted for high-carbon ferro-manganese in steel production, however, and since production of ferro-manganese totalled five to six million tons in 1973, it is possible that the additional supply a manganese metal from seabed mining would be absorbed at a price close to that of ferro-manganese. Current medium-carbon, ferro-manganese prices (mid-1974) are 31¢/lb., and in the absence of contrary evidence, we assume that this price can be maintained over time in real terms.

This analysis is predicated on the fact that most seabed mining ventures will not extract manganese and that Deepsea will extract manganese metal. If, however, more firms decided to extract manganese metal, the ferro-manganese market would also suffer from excess supply. Similarly, if firms decided to extract manganese ore, there would be some effect on the manganese ore market. There are indications, however, that manganese ore prices, current price 5¢/lb., and estimated 1985 price between 3¢/lb. and 5¢/lb., may not provide the necessary economic incentive sufficient for firms to extract manganese in its ore form [39].

As far as revenue forecasts for the extraction by Deepsea Ventures of minor amounts of vanadium, zinc and molybdenum, we assume in the absence of contrary indications that current market prices will be sustained in real terms until 1985. This is a simplifying assumption which appears warranted given the facts that only Deepsea will extract the trace metals and that the addition to total supply from these metals is small.

Table 6–3 reports forecasted metal prices and predicted revenue from seabed mining in 1985 in terms of constant 1974 dollars. As we have done throughout the analysis, we use 1974 as the standard of comparison. Prices will obviously rise in nominal terms over time; however, we are concerned with real prices in 1974. We thus adjust future prices by discounting them by the rate of inflation to obtain their real value in 1974. In cases where we have no indication that the current price of metal is not sustainable over time, we assume that the real price remains constant. Table 6–3 reports estimated revenue for seven-metal recovery operations ranging between $284 and $324 per dry metric ton of nodules. A similar calculation for a

Table 6–3. Price Forecasts and Revenue Projections

Metals	*Amount Contained per Metric Ton (lbs./ton)*	*Current Metal Price (1974 $/lb.)*	*1985 Price Forecast (1974 $/lb.)*	*1985 Revenue Projection (1974 $/ton)*
Nickel	33.6	1.85[1]	2.00–3.00[8]	67.20–100.8
Copper	26.9	.92/.65[2]	.75[9]	20.18
Cobalt	5.6	3.75[3]	2.00–3.00[10]	10.20–16.80
Subtotal for Three Metal Recovery				97.58–137.78
Manganese Metal	560.0	.42[4]	.31[11]	173.60
Vanadium	2.5	1.90[5]	1.90	4.75
Zinc	1.8	.40[6]	.40	.72
Molybdenum	3.4	2.30[7]	2.30	7.82
Total for Seven Metal Recovery				284.47–324.67

Table 6–3. Notes

1. Price of major producer cathode nickel as of May 15, 1974 as reported in *Metals Week*, V. 45, No. 40 (October 7, 1974).
2. Respective LME W-Bar settlement copper prices for January 1974 and October 1974 as reported in *Metals Week*, *supra.*
3. Price of shot/cathode/250 K.G. cobalt as of July 1974 as reported in *Metals Week*, *supra.*
4. Price of Manganese metal/regular as of July 1974 as reported in *Metals Week*, *supra.*
5. Price of U.S. pentoxide/fused vanadium as of August 1974 as reported in *Metals Week*, *supra.*
6. Price of LME-cash zinc as of October 1974 as reported in *Metals Week*, *supra.*
7. Price of climax concentrate molybdenum as of September 1974 as reported in *Metals Week*, *supra.*
8. As per estimated 1985 nickel price.
9. As per estimated 1985 copper price.
10. Assuming 1985 cobalt prices fall to those forecast for nickel.
11. Assuming that electrolytic manganese metal can successfully substitute for ferro-manganese (75 percent manganese contained) and given medium-carbon ferro-manganese prices as $.31 per pound as of September 1974. Sources: *Metals Week*, *supra.*

three-metal recovery operation yields a $98 and 138/dry metric ton estimate. Assuming 15 million tons of nodules are exploited in 1985, and only Deepsea extracts seven metals, the total revenue gained from deep seabed nodule mining in 1985 would total between $1.66 billion and $2.26 billion in 1974 constant dollars.

EXPLOITATION COSTS AND PROFITABILITY

Deep seabed mining for manganese nodules is extremely capital-intensive and involves high threshold costs. Estimates of the capital expenditures required to initiate mining at an annual level of one million dry tons per year range between $150 and $250 million [40]. Operating costs are reported to range between $20 million and $75 million per million tons, depending on the type of mining method and chemical extraction process [41]. Clearly there are very high, front-end capital investment requirements which only a few consortia from the industrialized nations have been able and willing to undertake. Inasmuch as expenditures to date have been primarily for exploration and research and development, not for the actual mining equipment, vessels or refining plants, all cost estimates are somewhat tentative. We nevertheless propose to examine the projected 1985 cost structures of Deepsea Ventures and Kennecott, the latter as an example of firms committed to three-metal recovery, and compare these costs to projected 1985 revenue. The analysis is undertaken in constant 1974 dollars.

Deepsea Ventures, which anticipates starting a one million ton per year nodule mining operation by 1980, has developed a hydroclorination process for extracting seven minerals from manganese nodules. The average cost of a processing plant to accomplish this was reported in mid-1973 to be approximately $35–50 per dry short ton [42]. Adjusting this figure for the average United States inflation rate of approximately 18 percent in the following eighteen months and further adjusting for metric tons, raises the average on-land operating expenses to $46–66 per ton as of the end of 1974 [43]. Direct operating costs of mining and transporting were given as $20–25 per dry short ton in 1973 and an adjusted 1974 figure per metric ton is $26–33 [44]. Thus total annual operating expenses for a one million dry metric ton operation are estimated to be in the range of $74–99 per ton. The reported capital investment required by Deepsea is on the order of $140–210 million for a one million ton annual production level [45].

Kennecott, a second producer likely to begin full scale nodule mining in the next decade, plans to recover and refine three million tons of dry nodules per year using a pyrometallurgical process to extract the three primary metals. The capital requirements for such a venture are reported to be approximately $300 million [46]. Total operating costs are projected to be $24–36 per dry short ton or perhaps $29–40 per dry metric ton in year-end 1974 prices [47]. This total cost figure is disaggregated into perhaps $8/ton for gathering and lifting the nodules, $4–9/ton for transportation costs, and $17–23/ton for processing costs [48]. The higher figures appear closer to the mark, at least in the first few years of production [49].

In any comparison of costs and returns one needs to consider the fact that expenditures take place over a period of time preceding the inception of production and the earning of returns. While the financial calculations have been adjusted for inflation, namely, both expenditure and income are calculated in 1974 constant dollars, capital costs figures must also be adjusted to reflect the lag between expenditures of capital and inception of returns. A crude estimate is that a firm hoping to begin nodule mining in 1980 would need to spend 2 percent of its total capital expenditures annually between 1970 and 1975 (viz., for exploration and R&D) and an additional 22 percent per year between 1976 and 1979 (viz., for mining equipment, vessels and a processing plant). Assuming that the opportunity cost of capital—the returns on alternative investments—were 8 percent per year, then an estimated capital expense of say $175 million over ten years, given the aforementioned pattern of expenditure, would really involve an expenditure of $225 million. Using a similar method to adjust the expenditure of $300 million over a ten-year period, say from 1972 until 1982, would yield a real expenditure level by 1982 of some $385 million in 1974 constant prices. Using the revenue data from Table 6–3 and the reported costs, 1985 profit levels and pre-tax profit rates as a proportion of invested capital can be calculated for three-metal recovery and seven-metal recovery operations. The data are reported in Table 6–4. The estimates, based on a number of projections subject to error, are that pre-tax profits would range between $185 million and $253 million per dry metric ton per year for a seven-metal recovery operation and between $58 million and $109 million per dry metric ton per year for a three-metal recovery operation. Pre-tax return on invested capital ranges between 82 percent and 112 percent for the former operation and between 45 percent and 85 percent for the latter.

It is revealing to compare these profit figures to the average mid-1974 to mid-1975 pre-tax profit of United States mining corporations, which was 27 percent per annum [50]. Manganese nodule

Table 6–4. 1985 Profitability Projections *(1974 U.S. m. $)*

	Seven Metals (1 m.m. tons)			*Three Metals (1 m.m. tons)*		
	Low	*Mean*	*High*	*Low*	*Mean*	*High*
Total Revenue	284	305	325	98	118	138
Total Costs (Including Debt Amortization)	72	86	99	29	35	40
Total Profit* (Pre-Tax)	185	219	253	58	83	109
Rate of Return on** Investment (%)	82	97	112	45	65	85

*Low profit is derived from low revenue and high cost options and vice versa for high profits.

**Assuming investments of $225 million and $385 million respectively and recalling that the latter expenditure is for a three-million-ton operation.

mining thus appears to be a highly profitable investment. It should be noted, however, that manganese nodule mining involves venture capital which is being invested in a new and highly speculative industry in terms of the technology needed, future costs and metal prices, and, most importantly, the international structure of the deep seabed mining regime. Were one to include the uncertainty surrounding many aspects of nodule mining, the "expected" returns would be far lower [51].

It is instructive to compare the preceding projections and estimates to those of industry and of the United Nations. A study by Moncrieff and Smale-Adams on the financial prospects of the Kennecott-led consortium is a good example of the former [52]. They report pre-tax profit rates for a three million ton per year operation of between 10 percent and 20 percent. These figures are based on the following assumptions: (1) mineral contents and recovery rates similar to those assumed in this analysis, resulting in almost identical metal outputs of nickel, copper and cobalt; (2) selling prices range from $1.50 to $2.00 per pound of nickel, $.60 to $1.00 per pound of copper, and $2.00 to $4.00 per pound for cobalt (viz., their cobalt estimates appear high and their nickel estimates are slightly lower than we would expect); (3) capital costs (unadjusted over time) are given between $250 million and $400 million, with an additional $60 to $150 million for exploration and R&D (viz., these figures appear unnecessarily high) and annual operating costs are given to be between $30 and $43 per short ton. Selecting the high revenue estimates and low cost estimates from this corporate report results in a more realistic pre-tax profit rate of approximately 50 percent. One could hardly expect so much investor excitement in manganese nodule mining if the pre-tax returns were between 10 percent and 20 percent!

The United Nations estimates that pre-tax returns on nodule mining investments could range between 43 percent and 109 percent for a seven-metal recovery operation and 54 percent to 94 percent for a three-metal recovery operation [53]. These estimates are based on projections similar to ours. The United Nations report calculates profits and returns on equity in the presence of an international mining tax, alternatively set equal to 30 percent on a royalty basis (a percentage of revenue) or equal to 50 percent of profits. In the presence of a 30 percent royalty, the profit rate of miners extracting three metals is estimated to range between 28 percent and 58 percent of total investment annually, and a similar measure applied to seven-metal recovery results in returns on investment of between 18 percent and 63 percent. With a 50 percent royalty, the rates of return are reported to be between 27 percent and 47 percent and between 22 percent and 55 percent, respectively [54].

The consensus is that there are substantial profits to be earned in manganese nodule mining. Compared to a pre-tax rate of return of 27 percent earned by United States mining firm's in 1974–75, a consensus of the estimates available shows manganese nodule mining averaging twice that amount. Although entry requires a high level of capital investment, the relatively low operating costs and the size of the deposits render large-scale manganese nodule mining a highly attractive investment. The key to its actual profitability, however, is the nature of the deep seabed mining regime—namely, the rules and regulations of mining in international waters. Are miners subject to international taxation and if so at what rate? What are the prospects for the security of the investment in terms of exclusive mining claims to mineral deposits in areas outside the limits of national jurisdiction? Will miners be subject to any regulations in terms of determining an efficient level of production or in terms of marketing their output? These crucial issues are examined in the following chapter.

EFFECTS OF DEEP SEABED MINING ON DEVELOPING COUNTRY MINERAL PRODUCERS

Some attempts have been made to answer the extremely important question of whether or not developing country land-based producers will be significantly affected by deep seabed mining [55]. Quite apart from the problem of forecasting, however, is the problem of defining exactly what constitutes lost revenue. Is it the difference between that amount of revenue an exporter might earn in 1985 in the absence of nodule mining and the amount actually earned in the presence of seabed mining? This measures the absolute difference in earnings at some future point in time. Or is it the difference between actual earnings in a milieu including seabed producers compared to the trend of past earnings? This would adjust an earnings difference by the average of past earnings or account for a historical trend line in earnings. Or perhaps one should include "losses" associated with a decision not to exploit newly discovered mineral reserves because of the threat of seabed competition. This would include the loss associated with the foregone opportunity to mine a deposit.

The objective here is to concentrate on actual losses in the export earnings of traditional land-based producers which are clearly ascribable to seabed mining. Those developing country producers are identified whose export earnings are substantially dependent on affected minerals and where reductions in earnings are significant in magnitude.

Setting such limitations involves a decision to exclude certain possible cases. It excludes the cases of mineral exporters whose economies are not crucially dependent on affected minerals, e.g., say affected mineral exports are less than 5 percent of total exports. (In these cases, it is to be expected that adjustment to a reduction in revenue from a minor export will be manageable.) It also excludes producers entering the market between 1976 and 1985, since those producers are well aware of the potential risks to be encountered as a result of seabed mining. And finally a decision to limit the inquiry into metal markets in which additional supplies from seabed sources are a significant proportion of total supply excludes cases in which a small addition to net supply (e.g., say 5 percent or less) can have significant effects on land-based mineral producers [56]. Additional mineral sources—land-based or from the seabed—can always affect producers, and the international community should not be expected to guarantee mineral export earnings [57]. What should concern the international community, however, is the future impact on traditional, undiversified LDC mineral exporters of an internationally agreed upon effort to extract minerals from a new, large source of supply. We examine the cases of copper, cobalt and manganese [58].

To begin with, let us examine the case of copper, the seabed mineral of most interest to LDCs because copper exports play a vital role in the economies of developing country producers, especially Zaire,

Zambia, Chile and Peru. Total copper exports from the largest six LDC producers were valued at $4.65 billion in 1974. Given the magnitude of developing country copper exports as sources of foreign exchange, slight price or output variations can result in sizeable differences in foreign exchange earnings. Calculations indicate that copper from seabed sources in 1985 would account for 1.9 percent of total world copper production. A recent UNCTAD study on the effect of deep seabed mining on the copper market concluded that such an increase in supply would be well below the average annual fluctuations of copper mining output in recent years [59]. Therefore, it is difficult to argue that seabed supply would add a significant new degree of uncertainty into the determination of market prices for copper. The United Nations Secretariat Report on the economic implications of deep seabed mining, prepared for the Law of the Sea Conference, concludes that "production from nodules is expected to have a very minor impact on copper markets by 1985" [60]. Any such prognosis, however, is sensitive to the base year and future year selected for analysis. Therefore, one cannot say that in the year 2000, for example, seabed mining might not be a major source of copper, with concomitant effects on land sources of supply. What can be said, though, is that knowing what we do now about first generation mining in 1985 and given reasonable assumptions, it is highly unlikely that land-based copper producers will be adversely affected to any appreciable degree by 1985 levels of deep seabed mining.

In the case of cobalt, however, there is reason to believe that land-based producers will suffer significant losses as a result of seabed production in 1985. Of the previously forecast 1985 consumption demand of 60,000 metric tons, 62.5 percent or 37,000 metric tons might be supplied from seabed sources. Since the demand for cobalt is fairly insensitive to price changes, an excess of cobalt supply would significantly reduce prices, at least to that of its closest substitute, nickel [61]. Should cobalt prices remain at their present real level in 1985 (i.e., $3.75/lb. in 1974 prices) in the absence of seabed mining, then LDC producers, maintaining their present 75 percent of the market, would earn $8400 per metric ton on 45,000 tons of cobalt exported. Should the cobalt price fall to that of nickel in 1985 as a consequence of manganese nodule mining, then assuming a medium nickel price of $2.50 per lb., LDC revenue would be $5600 per metric ton on 45,000 tons. Assuming all produced cobalt could be sold at $2.50 per lb., the net LDC revenue loss in this 1985 scenario would be $126.0 million, of which Zaire (assuming a constant market share) would lose $88.2 million [62].

Current LDC cobalt export earnings are $190 million per annum and in the absence of nodule mining these export revenues can be expected to increase in real terms to a level of $378 million per annum in 1985. Manganese nodule mining might under our previous price assumptions lower these earnings to $252 million in 1974 prices. Therefore land-based cobalt producers can expect a steady rise in cobalt export earnings until the early 1980s, when seabed production will dampen earnings growth. There is only one developing country, Zaire, which is heavily dependent—as of 1974 at least—on cobalt export earnings as a prime source of foreign exchange. (See Table 5–7.) If Zaire, as the major traditional cobalt producer, is to avoid a major decline in export earnings, it will become necessary to implement export diversification programs as one means of reducing the adverse effects of manganese nodule mining. To assist such a diversification effort, manganese nodule miners might be asked to contribute to some form of new compensatory facility [63]. A discussion of existing compensatory financing facilities and possible new facilities funded by revenue from manganese nodule mining is contained in the following section.

The case of manganese is a rather difficult one to examine, since present indications are that firms will not extract manganese ore from nodules but that if all firms did decide to do so, the manganese ore market would be affected to some extent. Were all firms to extract manganese ore, the net addition to total supply would be approximately 10 percent [64]. This increase in supply might exert some downward pressure on prices and might affect the export earnings of LDC land-based producers. As reported in Chapter 5, however, only Gabon is currently dependent on manganese ore exports for more than one percent of its total foreign exchange earnings, and Gabon's export profile was recently radically improved with the discovery of onshore and offshore oil deposits. Therefore, the outlook for Gabonese export earnings is rather optimistic with or without nodule mining.

The latest information available suggests that Deepsea Ventures will extract manganese in its metallic form; the Kennecott and INCO consortia have not indicated whether they will extract manganese ore, stockpile it or discard it. At first glance, the last course does not appear likely considering the high cost of recovering nodules from the deep ocean floor and the first course does not seem likely given the current and future projected low prices of manganese ore. The decision of whether to sell, discard, or stockpile tailings (i.e., mineral residue perhaps containing recoverable amounts of metals) depends on the costs of storage, costs of extraction, price of the metal and expectations of future metals prices and costs of extraction. In the absence of specific information, it is conceivable that while it may not be terribly profitable to extract manganese ore in 1985 (given present assumptions about the manganese ore market in 1985), the ore might be stockpiled for future use [65].

Estimates concerning the effects of manganese nodule mining on land-based producers of manganese ore are therefore rather uncertain. It does not appear, however, that the maximum 10 percent addition to the supply of manganese ore that occurs if all nodule miners were to extract manganese ore is likely to affect prices radically in 1985. To the extent that the ore is stockpiled, however, nodule mining firms will gradually be building stocks of manganese ore which by their very existence constitute a threat to the price of manganese ore past 1985. If all producers except Deepsea mine and stockpile manganese ore, the stockpile by 1990 would be sufficient to interfere substantially with the market. Either safeguards need to be built into international regulations on seabed mining or perhaps by that time international commodity agreements to stabilize prices and assure supplies over time will be prevalent and thus stockpiling would not be a potential price destabilizing factor [66].

COMPENSATORY FINANCING

Exporters of minerals are subject to fluctuations in earnings, and there is no reason to believe that these fluctuations will be reduced in a mining milieu including deep seabed mining. Indeed, in some cases, particularly cobalt, and possibly to some extent in nickel and manganese ore, deep seabed mining may result in a significant reduction in the export earnings of land-based mineral producers. One method of assisting countries whose export earnings are reduced by factors external to the producing country is through the International Monetary Fund's Compensatory Financing Facility, a source of financing which has been in operation since 1963 [67].

The Compensatory Financing Facility was recently liberalized to allow countries greater borrowing limits. Members of the IMF may borrow foreign exchange to cover an unanticipated export shortfall at interest rates varying between 4.0 and 6.0 for loans of one to five years in duration and at 6.5 percent per annum for loans in excess of five years. Except in major emergencies, a country cannot borrow more than 50 percent of its quota in any one year, subject to a total 75 percent of quota limit. Table 6–5 lists the present and revised quotas of major LDC mineral producers, the maximum they may borrow under compensatory financing and their current borrowings as

Table 6–5. IMF Compensatory Financing Facility Borrowing *(Millions SDRs)**

	IMF Quota	*Maximum Drawings (75% of Quota)*	*Current Drawings*	*Revised IMF Quota*	*Maximum Future Drawings (75% of Quota)*
Zaire	113.0	84.8	28.2	152.0	114.0
Zambia	76.0	57.0	19.0	141.0	105.8
Philippines	155.0	116.3	29.1	210.0	157.5
Chile	158.0	118.5	76.0	217.0	162.8
Peru	123.0	92.3	—	164.0	123.0
Brazil	440.0	330.0	—	665.0	498.8
Mexico	370.0	277.5	—	535.0	401.3
Ghana	87.0	62.3	—	106.0	79.5
Gabon	15.0	11.3	—	30.0	22.5
Morocco	113.0	84.8	—	150.0	112.5
India	940.0	705.0	62.0	1145.0	858.8
Indonesia	260.0	195.0	—	480.0	360.0
Dominican Republic	43.0	32.3	—	55.0	41.3

*The value of the SDR is a changing weighted average of sixteen major currencies. In recent experience, an SDR unit has tended to equal approximately $1.15–$1.20.

Source: I.M.F., *International Financial Statistics* (October 1975).

of October 1975. The revised quotas will most likely take effect in the next year, as countries contribute their increased quota assessments which were a result of the Sixth Annual Review of Quotas which took place at the beginning of 1976. It should be noted that the recent liberalization of the Compensatory Financing Facility not only raised drawing limits in any year from 25 percent of quota to 50 percent of quota (and the maximum total drawings from 50 percent of quota to 75 percent of quota), but also involved a more liberal interpretation of export shortfalls [68].

The Compensatory Financing Facility is an intermediate cost source of funds; interest rates are well below the commercial borrowing terms and in recent years have been below the average international rate of inflation, yet is it not highly concessional, like the soft-window terms of the International Development Association or the IMF's Trust Fund. Countries borrowing from the Facility must use the funds to meet unforeseen gaps in foreign exchange earnings, but must also couple this with export diversification if it appears that their traditional export products are facing difficulties. If not, the loan and accumulated interest cannot be repaid. In the case of developing country land-based producers of minerals competing with supply from the seabed, losses in 1985—the base year—may be only the tip of the iceberg. Future years will bring additional entrants into seabed mining, their number limited of course by the capital and technological requirements, and diversification becomes of paramount importance.

In the preceeding analysis, it was concluded that some developing country exporters might be slightly affected by seabed competition, as they would from any new land-based sources of supply. In most cases examined, the conclusion has been that either developing country mineral producers are not vitally dependent on earnings from a mineral source which will compete with a nodule-source of supply, or if the producer is vitally dependent, the price of that mineral is not expected to be significantly reduced as a result of manganese nodule mining. A significant exception, however, is the case of cobalt, where Zaire can expect to suffer a severe drop in export earnings as the price of cobalt falls [69]. Zaire's projected earnings loss in 1985 due to reduced cobalt prices, $88 million, is larger than its current maximum drawing potential under the Compensatory Financing Facility. While Zaire's borrowing ability should increase over time, it is unlikely that this source of financing is large enough to serve as a substitute for Zairian export reductions, because the facility is not designed to offset major, sustained reversals in export earnings.

Nevertheless, the IMF Facility can be extremely useful in providing short-term assistance for selected copper, manganese and nickel exporters affected by reduced export earnings. In many cases, the impact of nodule mining can be predicted as well and adjustments made to expand other exports. Indonesia in the case of nickel and Gabon in the case of manganese ore, are examples of developing countries, which, as oil producers, are in an excellent position to be able to adjust without much assistance to possible reduced foreign exchange earnings in minor exports. Similarly, Brazil and Mexico in the case of manganese ore should be able to offset any potential earnings reductions, which, as previously indicated, do not appear likely in 1985; these countries combine fairly high per capita income levels with a diversified export profile. While the major reductions in export earnings of copper and manganese ore producers are not anticipated, countries either at a lesser stage of development and/or more crucially dependent on particular mineral exports—e.g., Zambia, the Philippines, Chile and Peru in the case of copper or Ghana, Morocco and India in the case of manganese—might be potential borrowers of the Compensatory Financing Facility. Their total borrowing power should be in excess of $2 billion (viz., without India, this total drops to $1.5 billion) and this should be adequate to insure stable earnings, *ceteris paribus*, in 1985, given previous assumptions about the level of nodule mining and its effects on developing country producers of copper, nickel and manganese.

By way of summary, it should be noted that the real price of nickel is not expected to fall between 1974 and 1985 (viz., it is anticipated, however, that 1985 prices in the presence of nodule mining might be less than they would in the absence of nodule mining) and thus serious economic consequences for the two developing country minor nickel producers, the Dominican Republic (a new producer) and Indonesia (an oil producer), are not expected. We do not anticipate a significant effect of nodule mining on the copper market in 1985, because of the small increase in total copper supply. Still, since a few developing countries are heavily dependent on copper exports (especially Zambia, Zaire and Chile), the copper situation bears watching into the late 1980s as seabed mining increases. Here, certain protective measures seem to be in order, as we discuss below. In the case of manganese ore, we do not anticipate major effects, and the developing country producers are in our opinion sufficiently diversified in their exports to adjust to small variations in minor export's earnings. Still, lower income LDCs (e.g., Ghana, Morocco, India) may require some traditional financing from the IMF facility, should manganese ore prices fall in the late 1980s, and this will involve costs to the borrowing countries [70]. In the case of cobalt, Zaire will require assistance to offset earnings losses in the early 1980s.

To assist Zaire in diversifying its exports and to create a contingency ability to assist LDC copper, nickel and manganese producers in the next decade and beyond, we would propose that a diversification assistance contingency fund be established, utilizing a portion of the revenue to be derived from manganese nodule mining. The objective of such a fund would be to provide short- to medium-term assistance (i.e., one to seven years) for developing nations, supplemental to the IMF facility, for major and sustained export losses which result from nodule mining. Three types of assistance are possible: (1) anticipatory diversification assistance to countries like Zaire to stimulate export expansion in alternative commodities prior to the inception of deep seabed mining (and the anticipated losses in earnings) [71]; (2) subsidization of borrowing under the IMF facility for low-income LDCs, whose earnings are adversely affected by deep seabed mining; and (3) ex post diversification assistance to developing country mineral producers seriously affected by nodule mining and unable to finance diversification programs.

In the first case presented, such a fund has the advantage of attempting to encourage diversification prior to the accruing of actual losses. It would not offer compensation, but rather it would finance export diversification, for a limited period of time, through a program jointly developed between the developing country producer and an international development institution (e.g., World Bank). Since the expenditure of money would have to preceed actual nodule mining (and the collection of any international tax revenue), funds would have to be borrowed, perhaps from international taxes collected from the exploitation of offshore hydrocarbons [72]. Funds could either be concessional or interest free loans, depending on the LDC's income, the level of accretion of the contingency fund and the magnitude of applications for assistance.

In terms of subsidizing IMF borrowing, such a fund would insure that low-income (viz., perhaps GNP per capita below $350) developing country mineral producers, whose earnings are affected by nodule mining and who as a result seek short-term financial assistance, would receive an interest free IMF loan. In the third case, ex post diversification assistance, concessional or perhaps interest free loans would be extended, for a maximum of seven years from a revolving fund. Developing country land-based producers would be assured that if in 1985 or 1995, for example, deep seabed mining radically affected their export earnings, a specially designated additional source of financing would be available. A country could conceivably benefit from both subsidized short-term financing (to weather the gap in foreign exchange earnings) and ex post diversification assistance. Obviously, precise rules would need to be developed to coordinate among these schemes and with the IMF facility.

While the conclusion has been reached that a diversification fund created by the international community with seabed taxes would serve a useful purpose, it is necessary to retreat a step and examine the overall structure of an international deep seabed mining regime. Given some knowledge concerning the potential profitability of nodule mining and the possible effects thereof on individual developing country mineral producers, we now turn to the major issues of the Law of the Sea Conference's Committee I: what kind of an internationally agreed upon structure is negotiable for mineral resources in areas outside the limits of national coastal state jurisdiction? What should be the rules and regulations applying to manganese nodule mining? How should the gains to be derived from deep seabed mining be distributed? Chapter 7 examines these issues. In Chapter 7, we also re-examine the possible uses of international deep seabed mining taxes in relation to the losses which might accrue to developing country land-based mineral producers, and methods of funding diversification assistance schemes.

NOTES TO CHAPTER 6

1. The fact that all three ventures are consortia, with the inherent spreading of risk among many partners, is no accident. Seabed mining is not only risky, but also extremely expensive in terms of initial capitalization.

2. A detailed examination of the taxation issue is contained in Chapter 7.

3. Total supply need not be augmented if land and seabed producers collude to maintain a fixed supply or if traditional land-based producers adjust (downwards) their production levels once the magnitude of seabed production has been determined.

4. See for example, R. Branco, "The Tax Revenue Potential of Manganese Nodules," *Ocean Development and International Law Journal*, 1973, Vol. 1, No. 2, pp. 201–208. Note that tons referred to are metric tons unless otherwise specified. A metric ton is equivalent to 2240 lbs.

5. See United Nations A/Conf. 62/25 (May 22, 1974). While Summa Corporation (Hughes Tool) no longer appears seriously involved in deep ocean mining, other entrants are probable. Recent speculation for example, is on Lockheed Aircraft, a firm heavily involved in building the Hughes mining vessel, Glomar Explorer, and on CNEXO, a French firm.

6. Certainly such decisions are not irreversible, however, should the economics of extraction change over time; some firms may stockpile unused mineral material for later extraction.

7. A. Moncrieff and K. Smale-Adams, "The Economics of First Generation Manganese Nodule Operations," American Mining Congress (October, 1974).

8. See United Nations, A/AC 138/36 (May 1971). An updated UN document, A/Conf. 62/25 (May 1974) revises some recovery rates primarily raising the amount of recoverable copper; however, we believe the earlier figures are more representative of average assay values.

9. See note 6. Kennecott has recently revived its option to extract manganese ore from the nodules. See the testimony of Marne Dubs of Kennecott, "Current Developments in Deep Seabed Mining," Hearings, U.S. Senate Subcommittee on Minerals, Materials and Fuels, 94:1 (November 7, 1975) p. 15.

10. See note 3.

11. It should be recalled that if all mining companies extract manganese *ore*, supply would be increased by 3.75 million tons or 15.5 percent of current consumption.

12. K. Takeuchi, Unpublished Report on Copper Market Outlook (1971), World Bank. For a bibliography of copper models, see W.C. Labys (ed.), *Quantitative Models of Commodity Markets* (Ballinger, 1975).

13. *Ibid.* Also Varon and Takeuchi report that studies reveal the short-run price elasticity of copper demand to be .20 and the long-term elasticity to be 1.35. See B. Varon and K. Takeuchi, "Developing Countries and Non-Fuel Minerals," *Foreign Affairs*, Vol. 52, No. 2 (April 1974) pp. 498–510; and Charles River Associates Inc. "Economic Analysis of the Copper Industry," U.S. Department of Commerce, PB189–927, (March 1970).

14. U.S. Bureau of Mines. *op. cit.*

15. K. Takeuchi, *op. cit.*

16. World Bank sources. 1974 year end prices were 97¢/lb. compared to 57¢/lb. during most of 1975.

17. A/Conf. 62/25 (May 22, 1974) p. 41. The UN also forecasts a 1985 cobalt price of $2.00 in 1974 prices.

18. See Joseph R. Boldt, Jr., *The Winning of Nickel: Its Geology, Mining, and Extractive Metallurgy*, (Van Nostrand, 1967).

19. We assume nickel demand is price-inelastic.

20. A joint product is defined as a single technical process which yields more than one output.

21. By price inelasticity of demand, we mean that a percentage increase in price causes a very small but certainly less than proportionate reduction in the quantity demanded. Perfectly price inelastic demand means the quantity demanded is unresponsive to price changes.

22. Economic theory outlines a number of possible patterns of oligopolistic behavior including collusion, independent (Cournot-type) behavior, leader-follower (Stackelberg-type) output decisions, and minimax-maximin profit decisions. See Henderson and Quandt, *Microeconomic Theory*, Chapter 6.

23. Recent data place Canadian sulfide ore production casts at $.90/lb.

and Le Nickel lateritic ore production costs at \$1.75/lb. Source: *Metals Week*, March 18, 1974.

24. Researchers, including Varon and Adams for example, agree that nickel's special characteristics make it difficult and costly to substitute for in the short or even medium term. See B. Varon, "The Nickel Outlook Reassessed," Economic Staff Working Paper No. 135 (August 31, 1972) IBRD and F.G. Adams, "The Impact of Nickel Production from the Ocean Floor: An Econometric Appraisal" UNCTAD (20 November 1974).

25. F.G. Adams, *op. cit.*

26. B. Varon, *op. cit.*

27. This includes new or planned projects in the Philippines, Indonesia, Guatemala, Dominican Republic, Colombia, Venezuela, Brazil, Botswana and Burundi. See UN, "Economic Implications of Seabed Mineral Development in the International Area: Report of the Secretary General," A/CONF. 62/25 (22 May 1974), p. 34.

28. We hypothesize the following simultaneous system of supply and demand equations for nickel:

$$(1)\ Q^d = a_1 P + a_2 QS + a_3 t : a_1 = 0;\quad a_2 > 0;\quad a_3 \gtreqless 0;$$

$$(2)\ Q^s = b_1 P + b_2 PM + b_3 t : b_1 > 0;\quad b_2 < 0;\quad b_3 \gtreqless 0;$$

where Q^d = quantity of (market economies) nickel demanded.[a]

Q^s = quantity of (market economies) nickel supplied.

P = market price of nickel.[b]

QS = total quantity of (market economies) steel production.[c]

PM = price index of mining machinery and equipment.[d]

t = time variable.[e]

We hypothesize that the quantity demanded is invariant with respect to prices, that the quantity supplied is a positive function of price, and that there is a positive relationship between quantity demanded and steel production (viz., the major use of nickel), and that there is a negative relationship between quantity supplied and the price of mining equipment, a major cost item for capital-intensive nickel mining. The time variable in (1) reflects changes in prices of substitute metals and the time variable in (2) reflects changes in the technology of nickel mining.

We have a just-identified system of two simultaneous equations and two endogenous variables, P and $Q = Q^s = Q^d$ in equilibrium. The reduced forms are as follows:

$$(3)\ P = \frac{-a_2}{a_1 - b_1} QS + \frac{b_2}{a_1 - b_1} PM + \frac{(b_3 - a_3)}{a_1 - b_1} t$$

$$(4)\quad Q = \frac{a_2\, b_1}{a_1 - b_1}\, QS + \frac{a_1\, b_2}{a_1 - b_1}\, PM + \frac{(b_3\, b_1 - a_3\, a_1)}{a_1 - b_1}\, t$$

The ordinary least squares estimated equations for 1953–1973 are reported, (5) and (6):

$$(5)\quad P = \underset{(-3.95)}{-49239200} + \underset{(5.27)}{7342\, QS} + \underset{(4.05)}{70860\, PM} - \underset{(-3.14)}{195682\, t}$$

$$\bar{R}^2 = .90 \quad DW = 1.36 \quad n = 21$$

$$(6)\quad Q = \underset{(-1.0)^*}{-209.9} + \underset{(2.70)}{.64\, QS} + \underset{(1.35)^*}{4.02\, PM} + \underset{(.52)^*}{5.49\, t} \qquad \bar{R}^2 = .95 \quad DW = 1.66 \quad n = 21$$

where t statistics are in parentheses and are significant at a .01 level of significance unless asterisked.

$\bar{R}^2$ is the coefficient of determination; *DW* is the Durbin-Watson statistic; and *n* is the number of observations.

Our results indicate that the demand curve is infinitely price inelastic. The supply elasticity with respect to price is calculated to be .46, (evaluated at the mean) and 1.59 with respect to the price of equipment. The (mean) demand elasticity is .61 with respect to steel production.

[a]Data in (000) of short tons. Source: U.S. Bureau of Mines, "Minerals Yearbook," (1973).

[b]Data in U.S. dollars per (000) tons. Source: INCO of Canada, Ltd.

[c]Data in (000) of tons. Source: American Iron and Steel Institute, "Annual Statistical Report," (1973).

[d]Price index (1967 = 100). Source: "The U.S. Mining Market," *Engineering and Mining Journal and Coal Age*, Vol. 4 (1973) p. 406.

[e]$t = 1 \ldots 21$.

29. We realize that the assumption of perfect price inelasticity of demand in the longer run is not realistic; however, we have not seen any evidence concerning the development of lower-priced nickel substitutes in recent years and in the absence of a better assumption, we leave this rather strong assumption in force. It serves as a limiting case, namely, with an infinitely price inelastic demand curve, a rise in supply will exert the maximum downward effect on prices.

30. For simplicity we assume that the rate of inflation in mining equipment is equal to the average world inflation rate.

31. This is rather similar to the World Bank's estimated increase in world nickel consumption of 6 percent per year until 1985, B. Varon, *op. cit.*

32. If one posits a demand function of the form

$$(1)\quad Q^d = aP + D;\ a < 0$$

and a supply function of the form

$$(2)\quad Q^s = bP + C;\ b > 0$$

and evaluates price at equilibrium, $Q^s = Q^d$, we have

$$(3)\quad P = \frac{-C + D}{b - a}$$

which can be differentiated with respect to an exogenous shift in supply:

$$(4)\quad \frac{dP}{dC} = \frac{-1}{b - a}$$

In the case where the price elasticity of demand is zero, $dP/dC = -1/b$. In the case where the price elasticity of demand approaches infinity, dP/dC approaches zero. Similarly, as the price elasticity of supply approaches infinity, dP/dC approaches zero and when the price elasticity of supply is zero, $dP/dC = 1/a$. This can be confirmed graphically by referring to the figure below.

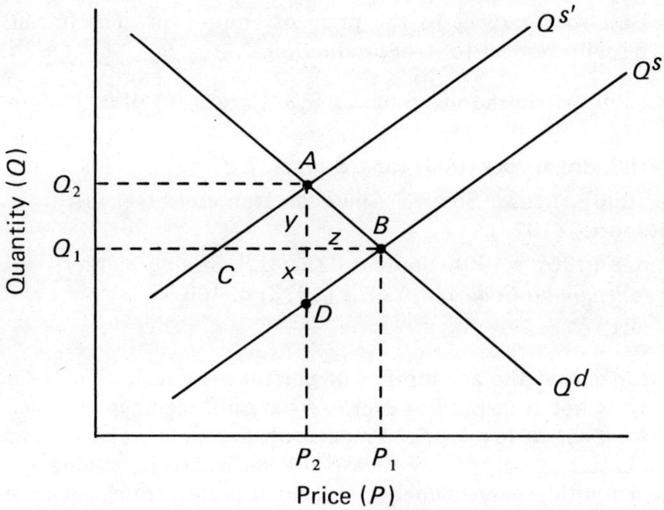

Plotting Q as a function of P and shifting from Q^s to $Q^{s'}$ to reflect an increase in supply, equilibrium moves from B to A. The shift in supply can be depicted by $x + y$ and the change in price can be depicted as z. The change in price for a change in supply is therefore equal to $-z/x+y$. By inspection, we know that $a = -y/z$ and $b = x/z$. Simple substitution yields $-z/x+y = -1/b-a$.

By using the definition of elasticities $\eta_{Q^d,P} = -(a)\ (P/Q^d)$ and $\eta_{Q^s,P} = (b)$ (P/Q) we can calculate the elasticity of price with respect to an exogenous shift in supply, $\eta_{P,C} = (dP/dC)\ (C/P)$.

$$(5)\quad \eta_{P,C} = \frac{-1}{\eta_{Q^s,P} + \eta_{Q^d,P}} \bullet \frac{C}{Q}$$

33. The more price inelastic demand is the more cost increases can be passed on to consumers of the final product, for example.

34. G. Adams, *op. cit.*, pp. 40–42.

35. *Ibid.*

36. *Ibid.*

37. World Bank data suggest that demand for manganese ore might rise by an average 3.7 percent per annum between 1973 and 1985, so that total production in 1985 is estimated to be 36.5 million m. tons.

38. Data for 1972. See N. Cornell, *op. cit.* and UNCTAD TB/B/C.1/ Conf. F/L.2. (February 1974).

39. World Bank price forecast.

40. See "Deep Ocean and Offshore Mining," for a summary of cost estimates.

41. *Ibid.*

42. Arnold J. Rothstein and Raymond Kaufman, Deep Sea Ventures Inc., "The Approaching Maturity of Deep Ocean Mining—The Pace Quickens," U.S. Senate Hearings (Subcommittee on Minerals, Materials and Fuels; Committee on Interior and Insular Affairs) *Mineral Resources of the Deep Seabed* (May 1973) pp. 201–222,

43, *Ibid.*

44. *Ibid.*

45. *Ibid.*

46. Marne Dubs, Kennecott Copper Corporation as reported in UN A/Conf. 62/25 (May 22, 1974) pp. 68–70, using an 18 percent inflation rate adjustment throughout.

47. *Ibid.*

48. *Ibid.*

49. The view of a Kennecott official in private conversation. An independent 1972 cost estimate of various deep-sea hydraulic lifts and processing methods concludes that total operating costs for a medium scale (viz., 1.5 million short ton operation), three-metal recovery project might be $29–33 per ton. The same source reports gathering and lifting costs to be $6.20–6.56/ton with approximately 10 percent of this representing direct operating expenses and 90 percent representing depreciation of capital equipment. See C. Glauss, "Theoretical and Experimental Investigations of Deep Sea Ocean Mining Systems and their Economic Evaluation," (Second International Ocean Development Conference, Tokyo, October 1972) pp. 1925–1955 as reported in UN A/Conf. 62/25 (May 22, 1974).

50. Federal Trade Commission, "Quarterly Financial Report for Manufacturing, Mining and Trade Corporations," (III, 1975) p. 14.

51. For a Theoretical discussion, see J. Tobin, "Liquidity Preference as Behavior Towards Risk," *Review of Economic Studies.* Vol. 25 (February 1958), pp. 65–86.

52. A.G. Moncrieff and K.B. Smale-Adams, *op. cit.*

53. UN A/Conf. 62/25 (22 May 1974).

54. Ibid., p. 76.

55. See, for example, UN "Economic Implications of Seabed Mineral Development in the International Area: Report of the Secretary General," A/Conf. 62/25 (22 May 1974).

56. These effects might be significant price reductions (and revenue reductions) due to the low price elasticity of demand for particular minerals or the exclusion of a high cost mineral producer from the market as prices fall and it is no longer profitable to produce.

57. African coffee producers were an unexpected new source of competition for traditional Latin American exporters, for example.

58. We do not examine the case of nickel in detail, since there are very few LDC nickel producers and their total output as a fraction of world production is approximately 7 percent; with the addition of Cuba, it is 12 percent. The two market-economy LDCs producing nickel are Indonesia, a major oil exporter, and the Dominican Republic; and in both cases production is just starting. We have no data on the value of Indonesian nickel exports, reported to be 16,000 metric tons in 1974. The Dominican Republic, which started nickel production as late as 1972, produced 31,200 tons in 1974, which was reportedly valued at $93 million. This constitutes 14.6 percent of total Dominican exports. Whereas the Dominican case warrants examination, the incipient nature of nickel mining and the uncertainty surrounding future production levels renders such an examination rather difficult.

59. UNCTAD, "The Effects of Possible Exploitation of the Seabed on the Earnings of Developing Countries from Copper Exports," TD/B/484 (28 May 1974), p. 7.

60. UN, A/Conf. 62/25 (22 May 1974), p. 36.

61. See UNCTAD, "Exploitation of the Mineral Resources of the Seabed Beyond National Jurisdiction: Issues of International Commodity Policy—A Case Study of Cobalt," TD/B/499/Add. 1 (June 26, 1973), p. 7 on cobalt demand. Also see UN, A/Conf. 62/25, where a $2.00/lb. 1985 cobalt price is forecast, and Cornell, *op. cit.*

62. An alternative scenario would be that in order to maintain the real price of cobalt equal to $3.75/lb. in 1974 prices, land-based producers voluntarily cut back production. Although this is unlikely in the absence of a producers' agreement, let us assume all land producers reduce output by 62.5 percent. LDC revenue would then be $148 million rather than the $378 million LDC land-based producers might earn without seabed competition. The net loss would be $230 million per year.

63. One could argue that Zaire has been gaining monopolistic profits over the

years by fixing the international cobalt price and thus is not a worthy candidate for compensation.

64. See p. 154.

65. Our conclusions are rather different from those presented by UNCTAD which reported a possible reduction in LDC manganese export earnings of as much as 30 percent from their normal levels in 1980. UNCTAD assumed that mining would begin in 1974 possibly reaching 10 million tons by 1980 and that all firms extract manganese ore, which is 30 percent of the nodule material. See UNCTAD, "The Effects of Production of Manganese from the Seabed, with Particular Reference to Effects on Developing Country Producers on Manganese Ore," TD/B/483 (23 April 1974). This analysis is in part based on an econometric model of the manganese ore industry which is highly suspect. See UNCTAD, TD/B/483/Add. 1 (23 April 1974).

66. See, for example, UNCTAD, "An Integrated Programme for Commodities," TD/B/C.1/185 (6 June 1975).

67. International Monetary Fund, *Articles of Agreement*, Washington, 1944, As amended by Decision No. 1477—(63/8), February 27, 1963.

68. On these issues, see T. Morrison and L. Perez, "Export Earnings Fluctuations and Economic Development," Agency for International Development Discussion Paper No. 32 (September 1975).

69. See p. 163.

70. Note that the real price of manganese ore is projected to fall over time, with or without nodule mining. See discussion on p. 155.

71. Certain threshold criteria would need to be established to determine access to an anticipatory facility, taking into account evaluations of potential losses in revenue resulting from deep seabed mining.

72. See p. 106 on taxation of offshore hydrocarbons by the international community and p. 201 for a discussion of deep seabed revenue generation.

Chapter 7

The Nature of a Deep Seabed Mining Regime

INTRODUCTION

The thorniest and yet perhaps most important issue to be negotiated at the U.N. Law of the Sea Conference is the determination of the nature of the deep seabed mining regime [1]. This issue has so polarized the industrialized and developing countries that the chances of negotiating an overall Seabed Treaty may be irreparably damaged. What is it about deep seabed mining that arouses passions, coalesces the developed and developing countries into strongly unified opposing groups and threatens the final outcome of the conference? One answer is that never before in international economic history has the international community laid common claim to resources outside the limits of national jurisdiction and then attempted to parcel out rights of access to and control over those resources. The schism between rich and poor nations is a reflection of the underlying disparity in national income levels which results in the fact that only a few industrialized nations possess the financial resources and advanced technology required to mine the ocean floor at great depths. The developing nations interpret the concept of "common heritage" to mean that the international community as a whole has a claim on seabed resources outside the limits of national jurisdiction, and thus they are committed to an exploitation system in which all nations directly share in the benefits to be derived from deep seabed mining [2]. While the discussions at present are in the context of manganese nodule mining, the developing world sees this concept applying to other ocean resources which may become significant at any time in the future. The industrialized nations, especially the United States where interest in nodule mining has been greatest, although having embraced the "common heritage of mankind" concept at the United Nations in 1970, do not feel that they have therefore altered their position on the right to unimpeded exploitation of deep seabed mineral deposits. A potentially new and abundant minerals source takes on increasing significance for the developing nations, many of which are mineral importers facing increasing demand for minerals, increasing costs of land-produced minerals, and attempts at cartelization by raw materials producers.

As was examined in Chapter 5, another factor which is central to the divergent interests of nations negotiating a deep seabed mining regime is that many developing nations are heavily dependent on mineral exports as sources of foreign exchange. A reduction in export earnings, perhaps caused by competition from nodule-source minerals, could seriously damage the growth prospects of some mineral exporting developing nations. While previous analyses demonstrate that the prospects of such major adverse effects of nodule mining are not great, the mineral-dependent developing nations are extremely concerned about such potential earnings losses and are therefore firmly committed to negotiating a system of deep seabed mining which does not threaten their mineral export earnings. Such a system might involve direct international management of deep seabed production levels, an interference in economic activity which is anathema to the "laissez faire" principles of economics espoused by many developed nations, especially the United States [3].

This chapter first examines the fundamental issues—and divergent interests and positions of developed and developing nations—surrounding a deep seabed mining regime. A second section presents some subjective views concerning the basic alternatives available to the international community and suggests some possible compromise outcomes. One compromise outcome is to combine economically efficient deep seabed production of minerals with a system of international taxation. Thus, the third section analyses alternative methods of international taxation, one method of redistributing the gains to be derived from manganese nodule mining. The final section examines the potential magnitude of tax revenues and analyzes possible uses of the revenue which would be generated from such a system of taxation.

THE BASIC ISSUES INVOLVED IN DEEP SEABED MINING

Among a host of important issues involved in determining an international deep seabed mining regime are the fundamental ones of who has access to seabed minerals and under what conditions. These questions fall inevitably back upon the shoulders of the ill-defined "common heritage" concept, unanimously agreed to in the United Nations in 1970 [4]. Some would argue, as does John Mero, the leading figure in manganese nodule mining over the past two decades, that common heritage is "a relatively banal use for a poetic and rather stirring slogan which might more eloquently and rationally be used to describe man's right to . . . breathe unpolluted air . . . but not to deep-sea nodules" [5]. Mero adds that a logical extension of the common heritage concept is that all nations should share in the ownership of Kuwait's or Saudi Arabia's oil [6]. The developing nations, however, argue that the common heritage concept is the main precept of any regime for the deep oceans and that common heritage implies common property ownership by the international community of all ocean resources outside the limits of national jurisdiction [7]. They maintain that the international community, as the custodians of the common heritage of mankind, should through an International Seabed Resource Authority (ISRA), control all aspects of deep seabed mining, including the power to determine who should gain access to deep seabed minerals. The industrialized countries, which originally accepted the common heritage concept, find it an increasingly burdensome concept, considering the degree of acceptance of a definition equating common heritage with common property and the extension by some of the common property concept to imply restricted individual access [8]. It is perhaps useful to reach beyond definitional questions to analyze the basic issue of access to deep seabed resources.

It is almost tautological to say that all nations should enjoy equal access to ocean resources which, it is agreed, are the common heritage of mankind. Having said that, however, one is confronted with the reality that not all nations—in fact, rather few—possess both the technology and the financial resources to engage in deep seabed mining. Therefore the adoption of a rule of equal access is devoid of any real positive economic significance for the majority of nations: most nations would not be able to benefit directly in deep seabed mining, as noted previously, and a particular group of nations might be expected to experience a negative economic impact from deep seabed mining. There appears to be a crucial difference between access to unpolluted air and to manganese nodules, to use Mero's example, in terms of who can potentially gain from that access.

It is also incorrect to argue, however, that the only effective method of insuring that developing nations participate in and benefit from deep seabed mining is to form a centralized body to collectively administer the common resource. There are economic benefits resulting from the efficient exploitation of resources; in particular, society benefits most from the exploitation of a resource if that resource is produced at the point at which the evaluation by society of the costs involved in producing an additional unit are just equal to the benefit accruing to society from the additional unit of output [9]. This condition prevails in a market in which all producers are perfectly competitive; namely, no single firm is large enough to affect the market price by changing its output, there is freedom of entry into the industry, and there is no external interference in the market. Underlying the industrialized nations' position is the view that deep seabed mineral resources should be freely exploited by competitive producers without constraints imposed by a Seabed Authority. Some would even argue that although developing nations may not directly gain benefits from the mining of nodules, they implicitly share in the benefits through normal trade in a competitive market milieu [10].

The international community is faced with two diametrically opposed positions concerning the nature of the deep seabed mining regime, the position of the developing nations, represented by the Group of Seventy-Seven (G–77), and the positions of the industrialized states, as seen in the negotiating proposals of the United States, the European Economic Community, Japan and the Soviet Union. The G–77 proposal concerning the structure of a deep seabed mining regime, first submitted at the Caracas session of the Law of the Sea Conference, calls for an International Seabed Resource Authority (ISRA) which would have wide-ranging powers over all aspects of deep seabed mining [11]. Specifically, the Authority, as the "agent for mankind as a whole," would either directly exploit the resources or enter into contracts with miners to undertake any phase of seabed mining [12]. Although the G–77 position provides for competitive selections among mining applicants, there are provisions that the Authority shall prescribe applicant qualifications and take into "special account the need for the widest possible participation of developing countries, particularly the land-locked among them" [13]. The United States proposal, tabled in Caracas in 1974, would allow the ISRA to enter into mining contracts with any party sponsored by a signatory to the Seabed Treaty on an unlimited, "first-come, first-served" basis with competitive bidding for claims filed coincidently by two or more parties [14].

The G–77 proposal would create a potential seabed monopoly, able to discriminate among mining applicants, and able to reserve choice mining tracts for its own use or for future use by developing countries. From the perspective of economic efficiency, such actions are likely to result in a misallocation of resources, since society would be using more resources than necessary to obtain a certain level of output [15]. The industrialized nations thus argue that discretionary access is inefficient and that a "first-come, first-served" system of access (perhaps combined with some form of international notification or registry) would insure timely and efficient exploitation of the deep seabed [16].

While the competitive free market approach to deep seabed mining can be most efficient, it need not be, however. If, for example, there are a limited number of mining firms who would control deep seabed output and would exact abnormal profits from consumers, the system cannot be described as efficient in the overall sense of efficiency [17]. An economically efficient system includes optimizing behavior by all actors in an economy—producers and consumers [18]. Whereas the developed nations urge the international community to adopt a deep seabed mining system which is economically efficient, they refer only to production efficiency, ignoring the distribution of benefits issue. The developing nations, on the other hand, are concerned with distribution of benefits and are willing to sacrifice efficient production. An economically preferred exploitation system would combine production efficiency with distributional efficiency or equity—a theme to which we return later.

Let us examine some of the specific proposals tabled by the Group of Seventy-Seven and the United States concerning the establishment of a deep seabed mining regime. The Group of Seventy-Seven proposes to create a strong International Seabed Resource Authority to regulate deep seabed mining [19]. Inasmuch as manganese nodule mining is not expected to begin until the turn of the decade, and many uncertainties remain, the G–77 proposal would limit the scope of current rules and regulations pertaining to deep seabed mining and would allow the ISRA the flexibility and powers to enact more specific rules and regulations at some time in the future. The ISRA would be the exclusive resource manager for the deep seabed [20].

The United States Caracas proposal seeks to establish a workable international deep seabed mining regime with a minimum of discretionary power vested in the Authority. The United States draft, which would specify, in a Seabed Treaty, the rules and regulations of deep seabed mining for decades to come, was characterized as a throwback to the lax land mining codes of the nineteenth century by the Norwegian delegate to the Law of the Sea Conference, Minister Evensen [21]. The United States proposal would grant individual deep seabed miners mine sites up to 30,000 square kilometers in size for up 55 years with rules specifying mining conduct at the outset and with limited powers of the Authority to alter those rules [22]. The position of the industrialized nations is that since manganese nodule mining involves the investment of large amounts of capital, the mining companies must be assured that the rules will remain fixed over the life of a contract with the Seabed Authority [23]. If the Authority retains the power to radically alter existing contracts or directly interfere in production decisions, mining firms could be forced to absorb sizeable losses at the hands of a capricious Authority. Given the polemics of international negotiations in U.N. fora, one must be concerned with the prospect of so large, vocal and diverse a body deciding on the economic issues of deep seabed mining on a day-to-day operational basis [24].

By way of summary, the developing nations are seeking to establish international control over deep seabed mineral resources by the creation of an International Seabed Resource Authority with substantial power. This desire to control deep seabed mining may be derived from the notion of common heritage or it may stem from the wish to extract the economic rent which is involved in deep seabed mining or it may be a desire to protect developing country land-based producers or it may be some combination of all three [25]. The Group of Seventy-Seven position does clearly indicate, however, that the developing nations are not going to sanction deep seabed mining internationally in areas outside the limits of national jurisdiction unless they directly share in the benefits to be derived from the deep seabed. The industrialized states, whose firms are anxious to begin deep seabed mineral exploitation, are determined to gain unimpeded access to seabed resources and are committed to the establishment of a deep seabed mining regime which is favorable to their interests. Given the ideological nature of the deep seabed debate in the LOS countries to date, it is not clear what developed nations would be willing to offer in the way of financial participation to the Third World nations in exchange for such access. In a very real sense, deep seabed discussions have been relegated to a backseat in the negotiations concerning national coastal state jurisdiction over offshore areas, rich in hydrocarbons and living resources. There has been little true negotiating and this dearth of progress has prompted some to think that the entire conference will fail to meet its objectives [26]. The following section outlines a number of ideas which might form the basis for a negotiated solution.

A FRAMEWORK FOR A DEEP SEABED MINING REGIME

The primary economic interest of the developed nations is to secure for themselves unimpeded access to deep seabed mineral deposits. In order to justify the substantial threshold capital investment necessary to enter the manganese nodule mining field, firms require assurances concerning the integrity of their investment. If, as is likely, most nodule deposits of interest to investors lie in international areas outside coastal state jurisdictions, and if mining is to be internationally regulated by a Resource Authority, then firms need assurances of competitive access to mine deposits and non-interference in production decisions. In the absence of such guarantees, the pace of deep seabed mineral exploitation will undoubtedly be retarded, reducing the possible gains to be derived from mining for all nations [27].

The primary economic interest of the developing nations is to appropriate a share of the benefits to be derived from deep seabed mining. A secondary, yet perhaps equally important interest, is the stability of the economies of the mineral exporting developing nations [28]. Developing countries maintain that it is not in their economic interests to agree to unlimited, free access to common resources, since they will be effectively excluded from sharing in gains of deep seabed mining, but will bear the brunt of possible losses.

The basic alternatives for a deep seabed mining regime which are available to the international community range from no international control to absolute international control over deep ocean resources. There are some who argue that there is no need to create an international regime to oversee deep seabed mining. They maintain that the lack of well-defined property rights will neither impede investment nor adversely affect production in the deep oceans [29]. It is debatable, however, that lack of an international structure for deep seabed mining will not affect investment or production. On the one hand, at present, the number of prime mine sites far exceeds the potential number of miners, so that firms need not be concerned about "claim jumping." Still firms expend large sums of money to explore alternative mining sites and develop equipment suited to a particular prime site [30]. As the number of possible entrants increases, the temptation to infringe on an "unofficially claimed" site increases. It is thus conceivable that conflicts similar to those now surrounding international fishing would develop in deep ocean mining. In view of the possible costs involved in a system which has the potential for anarchy, most seabed mining firms have stated a preference for a fair, internationally structured seabed regime, if such a regime can be fairly created [31]. Given the pace and direction of the recent Law of the Sea deliberations, however, many of the same firms might now prefer definite national mining legislation to further prolonged negotiations which might eventually result in an unfavorable investment climate [32].

There are a number of difficulties with the unilateral approach to deep seabed mining, quite apart from the political dilemma which would face a nation contemplating withdrawing from the Law of the Sea negotiations. Unilateral action would permit individual states to appropriate portions of the common heritage without regard to sharing benefits with those economies which do not directly participate in exploiting the common resource [33]. Such action fails to come to grips with the international nature of the deep oceans and sanctions the carving up of the deep seabed by a few industrialized nations in a manner similar to the appropriation of parts of Africa in the nineteenth century. To pass unilateral legislation regulating deep seabed mining is clearly in violation of the unanimously adopted United Nations Declaration of Principles, which states that "the area shall not be subject to appropriations by any means by States . . ." and that "no State . . . shall claim, exercise or acquire rights with respect to the area or its resources incompatible with the international regime to be established and the principles of this Declaration" [34]. The developed nations have become so disillusioned with the deep seabed negotiations, however, that the United States, for example, has stated its intention to mine the deep seabed, preferably but not necessarily within the content of an international agreement [35].

There are a variety of alternative multilateral approaches to international regulation of the deep seabed. These alternatives can be analyzed by examining the system of production which would be established, the system of distribution of gains which would result and the powers and the degree of flexibility in exercising those powers which would be vested in a Seabed Resource Authority.

At one end of the spectrum are those who wish to establish a free-market system of production with limited or no sharing of gains with the international community within the construct of an ISRA whose powers and functions are clearly and irreversibly determined at the outset [36]. Those committed to an efficient production system envisage an international organ which would either register claims or perhaps license vessels, but would not have any powers to discriminate among miners, affect production decisions, tax proceeds or otherwise interfere in the free market. There are a number of arguments in favor of such a system of deep seabed mining: (1) access would be non-discriminatory among firms from different nations; (2) production would be efficient insofar as firms would be able to control mining operations without interference; (3) deep seabed mining would be encourated, insofar as firms would be more willing to invest the large sums of money needed to initiate mining projects under a system of fixed rules and regulations; and (4) the international community—and particularly industrialized states—would be assured of access to new sources of certain minerals and would not be subject to a potentially capricious or ineffective authority. Arguments against such a regime for the deep seabed would include the following: (1) most nations, especially the LDCs, would effectively be excluded from sharing in the benefits of deep seabed mining; (2) the principle of the common heritage would be violated as mining firms would be able unilaterally to appropriate common resources without compensation; (3) ISRA could do little to protect society from the future actions of seabed miners in the field of pollution, for example, since its powers would be carefully circumscribed [37]; and (4) no provisions would be made for the potential losses by developing country land-based mineral exporters. The registry scheme obviously serves the interests of the industrialized states but does not adequately address the needs of developing nations.

At the opposite end of the international spectrum are those favoring an ISRA which is structured along the lines of the Group of Seventy-Seven position previously outlined, in which the Authority is less concerned about production efficiency than about the sharing of benefits from nodule mining. The Authority, as envisaged by the Group of Seventy-Seven, would be able to exploit deep seabed resources directly in competition with (or to the exclusion of) private firms. The advantages of such a regime for the deep seabed are said to be as follows: (1) the international community would exercise control over common resources in keeping with the United Nations Declaration of Principles and would thus directly share in the benefits of deep seabed mining; (2) the Authority would not be forced to operate for decades to come within the confines of a structure agreed to in 1976 in the absence of any real data on the nature of deep seabed mining, but rather would be able to evolve policies relevant to the problems as posed in the future. Those opposing such centralized resource management would argue that: (1) in the absence of predetermined and immutable rules and regulations, deep seabed mining will be significantly retarded; (2) market interference by ISRA—either directly in controlling output or marketing, or indirectly by favoring certain producers including itself in directly exploiting the seabed—is inefficient, potentially discriminatory, and not in the best interests of the world's consumers of minerals; (3) a potential seabed monopoly—whose objectives would be unclear and whose day-to-day operations would be controlled by perhaps as many as 140 nations—is unlikely to develop the deep seabed satisfactorily [38].

There is obviously a large and inviting middle ground to be examined. In such an examination, it is less useful—and quite possibly futile—to attempt actually to propose a structure for ISRA which could be acceptable to both developed and developing countries, than to attempt to outline the relevant economic characteristics of such a deep seabed mining regime. Our objective is to sketch the basis for a system which combines an efficient system of production with an equitable system of distribution of gains.

In espousing an efficiency-cum-equity system, one can argue that it is in society's interest to exploit deep seabed resources (e.g., manganese nodules) efficiently, as would occur in a competitive system but would not occur with a centralized exploitation system. In any international regime for the deep seabed which allows free market access to deposits and uncontrolled production, safeguards need to be included so that private firms do not impose any costs on society other than those of resource exploitation; for example, in the area of pollution control, rigorous international standards must be applied if the ocean floors are not to be permanently damaged, the costs of such damage being borne by society [39]. A system of efficient resource exploitation must be combined with a distribution system in which all nations share in the gains of deep seabed mining [40]. Just how such shares are derived is a very challenging question.

One method of sharing is to calculate the return due society for the use of its resources by ascertaining the resource rent of a deposit [41]. While the direct way of extracting the resource rent would be to auction deep seabed mineral deposits competitively, there is a problem with so doing; namely, the market for manganese nodules (vis., nodules are the example used although in principle these issues could apply to other seabed minerals) is not well developed, since there are few buyers and many sites. The value of nodules on a given site is hard to ascertain because mining has not yet begun, and thus information concerning the returns to nodule mining is rather tentative, and also because a single mine site may be exploited for as long as 40 or 50 years.

An alternative method of exacting the resource rent due society would be to assume that all factors of production, including entrepreneurship, receive competitive returns and then tax the profits in excess of normal rates of return. The difficulty with this approach is that the number of mining firms in the early years of exploitation—because of the technological and financial requirements and perhaps for decades to come—will probably be rather limited; thus firms can be expected to earn abnormal profits. Even if more advanced developing nations (perhaps with OPEC partners) were to enter the nodule mining field, it is probable that the number of market entrants would be sufficiently small so that firms would earn abnormal profits. Thus to tax all excess profits as a residual would most likely garner more than the amount of resource rent and include much of the economic rent normally collected by the entrepreneur [42].

As part of a compromise solution, the industrialized nations could be willing to share not only the resource rent, but also a portion of the economic rent with developing nations in order to secure unimpeded access to deep seabed minerals. If there are substantial profits to be earned in deep seabed mining, then a sharing of the total gains to be derived seems an acceptable alternative to either centralized exploitation or exclusion of LDCs. As Logue and Sweeney state:

> The common heritage concept suggests a general allocation to all nations of the world. Concern with international inequalities in levels of income and development may suggest that higher income, developed countries forego a portion of their share of this revenue in order to increase the funds available to lower income countries [43].

As a *quid pro quo* for a share in the benefits to be derived from deep seabed mining, the international community would need to create a stable investment climate, conducive to stimulating deep seabed mineral production. A stable investment climate could be created around an international authority whose mandate was to ensure the efficient and timely exploitation of deep ocean resources. ISRA would be a resource developer and its mandate would be clear.

A major problem, however, is the extent to which the Authority could alter the rules and regulations of investment mining—the climate—over time. At the Caracas session of the LOS deliberations, delegates from Jamaica and Trinidad and Tobago outlined similar sets of conceptual approaches to a structure for the Authority: (a) a mining code detailing in advance all the rules and regulations of mining with little or no discretion vested in the Authority and little flexibility for changes in the basic conditions of mining, except through cumbersome international legislative processes; (b) a strong Authority which could evolve freely over time and adjust the basic conditions of exploitation as circumstances warrant it and (c) an establishment of fundamental norms and unalterable basic conditions which would provide a framework within which the Authority would have the power to adjust to future economic realities.

Certainly, the last option is the most promising in terms of finding a mutuality of interests between industrialized and developing nations. If both groups of countries were committed to an efficiency-cum-equity solution, then one might envision the following scenario, for example: ISRA, as a resource developer, would lease tracts to miners, sponsored by states, subject to agreed-upon regulations. It would receive a share in the gains of deep seabed mineral development, the exact nature of the share to be examined in the subsequent section. The system would be nondiscriminatory; ISRA would supervise mining operations and strictly enforce the regulations, especially on maritime pollution. Miners would be required to supply the Authority with technical information as needed and would be asked to participate in the transfer of technology to LDCs, where appropriate, and in the training of LDC personnel. Safeguards for the export earnings of developing country mineral producers would be established. The Authority could not itself participate in mining ventures; however, it could offer economic incentives (perhaps with reduced taxation levels) to joint ventures between industrialized and developing countries. To anticipate the possible development of ocean minerals other than manganese nodules, the Authority should be given some flexibility in terms of its powers and functions. Some flexibility would also be granted, subject to appropriate legislative procedures, to alter certain regulations over time, with appropriate lead-time to alert miners to change and allow them to exercise alternative rights [44]. Of course, the specifics of these items would need to be negotiated within the framework of an efficiency of production—equity of distribution solution.

It is useful to delve more extensively into the issue of renegotiation of contracts for the exploitation of raw materials, since it is at the heart of the flexibility problem. In a sense, the issue is not unique to deep seabed mining; it is a problem for owners of all natural resources. There is a worthwhile analogy to direct foreign investment by industrialized nations in the extractive industries in developing nations, for example. There is a great deal of pressure on the resource owner to offer concessions at first and later, once the equipment is *in situ*, to attempt to alter the contract or perhaps even expropriate the investment. There is also an incentive, in the presence of such a threat, for the investor to exploit the resource at a rate greater than would be optimal. It would be preferable if resource owners and investors bargained on the basis not only of current strengths but on the basis of perceived future relative bargaining powers.

In order to deal with the problem of uncertainty inherent in nodule mining, which renders taxation rates either inadequate to capture economic rent or excessive enough to limit involvement, contracts could perhaps contain renegotiation clauses after a fixed period of time. One possible method of both altering contracts to reflect changes in revenue and costs over time and still guaranteeing firms a respectable return has been suggested by Garnaut and Ross, whose proposal was referred to earlier in terms of hydrocarbons [45]. Garnaut and Ross propose a "resource rent tax," a profit tax that begins to be collected when a certain threshold rate of return on investment has been realized and is automatically adjustable over time. The purpose of the tax is to assure the investor a sufficient return to stimulate new investment, yet to allow the resource owner to capture the remainder of the economic rent. It is this kind of symbiotic relationship which is required for a successful deep seabed mining regime—a regime that will stimulate deep seabed mineral development yet will also garner a portion of the gains for distribution to developing nations. The following section analyzes methods by which these gains may be shared.

INTERNATIONAL TAXATION OF DEEP SEABED MINING

There are a number of important issues to be examined in the construction of a system of international taxation of deep seabed mining [46]. Among them are the selection of a type of tax, the tax base, the tax rate and the compatibility of deep seabed and land taxes for particular minerals. The type of nodule tax is important since it can have varying effects on consumers and producers of the metals—in terms of output, price, incidence (viz., who pays the tax) and revenue generation. The question of tax base centers on the uncertainty surrounding the value of the nodules, *per se*, or alternative methods of imputing their value. The tax rate is of importance insofar as alternative rates can be combined with varying tax types or tax bases to yield identical revenue levels. The issue of tax harmonization refers to concerns that disparate tax levels on land-based and seabed sources may distort production away from the lowest-cost resource, which is sub-optimal from society's point of view.

In selecting a type of tax, one should be guided by the objective of the tax and the extent to which it meets that objective, as well as its effect on producers and consumers of the commodity in question. A tax on a natural resource can have a variety of objectives; for example, it can seek to stimulate a socially optimal level of production. This argument is based on the contention that the resource is scarce and therefore a tradeoff exists between present and future levels of exploitation. Individual firms might tend to exploit the resource at a faster rate over time than might be socially optimal, since firms are subject to the uncertainties of inter-societal transfers to which societal resource exploitation would not be subject. In other words, being subject to uncertainty, firms may attempt to gain higher expected profits by producing more now and conserving less. Taxes may thus be imposed to increase production costs and thereby reduce development to a level closer to the social optimum [47].

Another objective of taxation may be to capture the economic rent accruing to mineral producers in order to distribute these economic gains to others. This line of reasoning follows the argument that those appropriating a common resource should compensate those unable to share directly in the gains of exploitation [48]. It is not likely that developing countries will directly participate in seabed mineral exploitation; in addition, given the imperfection of markets facing developing nations and the small direct levels of consumptions of affected metals by developing countries, it is also unlikely that increased mineral supplies will cause price reductions to trickle down to LDCs [49]. Some method of international taxation appears to be the only effective method of directly including the developing nations in the benefits to be derived from deep seabed mining.

A third objective in taxing the exploitation of a natural resource could be to force the producer to internalize an external cost of production, such as pollution. The purpose of a pollution tax would be to raise revenue to compensate society for the costs imposed by the firm's actions and in so doing to induce the firm to produce a level of output at which the marginal social cost to society of a unit of output just equals the benefit society receives from an additional unit of output.

To argue that a conservation tax on manganese nodule production is needed would be inaccurate, since the quantity of available nodules is virtually unlimited [50]. To argue for a pollution tax would be premature, especially since it is hoped that stringent ISRA regulations will force firms to internalize the costs of pollution. An international deep seabed mining tax is necessary, however, to redistribute economic gains, the question being what type of tax would generate revenue without impeding the pace of mineral development, while meeting this aforementioned objective.

The major taxation options for the deep seabed consist of various forms of lump-sum taxes, production sharing, profits taxes and royalties [51]. By lump-sum taxes is meant taxes, the costs of which are additions to a firm's fixed costs. These might include license fees or rental fees, or bonus bidding, a payment offered (perhaps in a sealed bid type of auction) prior to the inception of production to secure access to the deposit. As has been discussed, it is unlikely that the international community would be able to secure the economic rent—or a substantial portion thereof—through fixed fees or bids [52]. One of the major reasons for this, apart from possible market imperfections, is that manganese nodule mining is a highly risky venture; the real value of access to a mining site in the years 1985–2015, for example, cannot be accurately judged, and thus the risks are reflected in the lower bids or fees offered than would be offered in a world of certainty [53]. Fixed fees would also be highly distortive if applied evenly to deposits of unequal economic value and would not serve to capture the excess gains to be derived from nodule mining. Bonus bidding for long-term access to deposits does not appear to serve the interests of the common resource owners either; in fact, inasmuch as it places additional front-end costs on the investor, it may not serve the miner's interests as well, as it renders entry into the industry costlier and potentially discourages deep seabed investment [54]. The main advantages of a lump-sum tax is that it can be easily applied to all countries in a non-discriminatory fashion and that it is compatible with a type of authority which exercises a limited direct involvement in deep seabed mining [55]. In summary, however, lump-sum taxes are not likely to garner the economic rent involved in deep seabed mining for the international community.

A second possible method of international taxation of deep seabed mining is a tax in kind—the sharing of a portion of production. This alternative might realistically be compatible with a decision to involve ISRA directly in deep seabed mining. Quite apart from the advantages and disadvantages of such a decision, production sharing involves a number of serious problems. While it would directly link the sharing mechanism with current price and output levels, since there is no market for nodules, *per se*, the Authority would either be forced to sell the nodules back to metals processors or to process the nodules themselves in competition with private firms. It is not clear whether the ISRA would attempt to sell the nodules—and there would be few buyers except the original miners—to maximize profits, or attempt to invest in processing plants, which is rather costly and requires a minimum scale of operations. The Authority would be placed in the position of being a major minerals seller, and possibly speculator, and perhaps significantly influencing particular metals markets with its actions. Another problem would be that the international community would be receiving a portion of the manganese nodules but would not be assured that this was equivalent to the economic rent [56]. Thus one would need to calculate the rent element (including resource rent and the economic rent) and attempt to match that up with a share of production. If this cumbersome mechanism were combined with a "buy-back" arrangement with miners, it would not appear preferable to the outright taxation of profits; and if it weren't, it would involve the Authority in the metal processing industry, which hardly seems a wise investment because of the technological and financial requirements.

A profits tax is a tax of net income like the corporate national income taxes in many countries which is most able to approximate the taxing of economic rent [57]. The tax varies directly with the firm's ability to pay and allows the international community's share to fluctuate with market conditions—more in good and less in lean years. A profits tax automatically adjusts the share to changes in metals prices and changes in technology and other factors influencing mining costs, all of which are expected to be quite variable over time. As an instrument of tax generation, it is efficient inasmuch as it does not affect the firm's output decision.

The major difficulty of a profits tax on manganese nodules is that profits, the tax base, are a result not only of the efficiency of a firm but also of its business practices and of the tax laws in the firm's country of origin. Since an additional ISRA tax administration can be expected to be cumbersome and costly, sponsoring states would be likely monitors of miners' income and the incentive to minimize international tax obligations would be high. There are a variety of methods available to "hide" profits, including, for example, intracorporate transactions, depreciation methods and reinvestment of earnings. In addition, firms from socialist economies operate on completely different standards, and the notion of profit is not strictly analogous [58]. A profits tax could thus be ineffective, highly discriminatory among firms and particularly among countries and extremely difficult to administer.

A royalty refers to a production tax levied upon the product at a specific stage of production. If one were to attempt to tax the value of manganese nodules directly, at the mine site for instance, it would be terribly difficult to set an appropriate royality on production since the value of nodules at the mine site is unknown. If one wishes, on the other hand, to tax the value of the refined metals extracted from manganese nodules, then such an *ad valorem* sales tax is easy to devise. A royalty could be levied on the basis of the average assayed value of contained metals, evaluated at final market prices. Such an *ad valorem* sales tax would not discriminate among firms recovering different combinations or future combinations of metals (i.e., three- versus seven-metal recovery, for example) or among firms with differing production methods, recovery rates or cost accounting systems. Market prices of metals are readily available, and the tax generated would directly reflect the current economic gains of nodule mining.

One of the drawbacks of a sales tax is that it is an addition to the marginal cost structure of the mineral exploiter and thus will reduce output from the pre-tax level. Several commentators are concerned about sub-optimal production of mineral deposits [59]. Another concern is the fact that all or part of the tax cost can be passed on to consumers of final metals if metal markets are not perfectly competitive [60]. In such cases, firms could still be earning abnormal profits and metals consumers would be funding all or part of the share of the international community. While this still serves the purpose of redistribution of gains (as long as consumers are not the non-participants in mining), there may be reasons why this is not acceptable to the international community [61].

A final option is the amalgamation of a royalty and a profits tax a so-called two-tier system [62]. Such a scheme would allow the international tax obligation of miners to be established through a royalty on sales, while the actual collection of revenue to pay the tax obligation would be through a profits tax. The royalty-profits combination would combine the advantage of a direct tax on the value of the metals contained in manganese nodules with a non-distortive method of collection. Imagine that by international agreement, firms are required to pay the Authority a tax equal to a hypothetical 20 percent of the market value of metals extracted from manganese nodules. A firm mines a million tons of nodules with an average assayed mineral content of 1.5 percent nickel, 1.2 percent copper and .25 percent cobalt. If current prices were $2.00 per pound of nickel, $.75 per pound of copper and $2.00 per pound of cobalt, then the value of extracted metals would be $98 million. The tax obligation on the firm would total $19.6 million. Thus, if the state is market-oriented in its economic philosophy, it may choose either to impose a profits tax on the mining firms or to allocate a portion of existing profits tax collections from the firm for transfer to the Authority. An interesting question arises if the hypothetical firms' profits do not total $19.6 million. Then the state is responsible to pay the tax and can initiate whatever domestic proceedings it wishes to recoup—or the firm will be in violation of its contract with the Authority and lose its license. This is a feature of a royalty or fixed cost obligation; namely, it can result in a no profit situation with the firm either borrowing, ceasing to exist or being subsidized by its sponsoring state.

In evaluating the relative merits of a royalty, profits tax and combination thereof, one should examine the method by which the rate of taxation might be set and the extent to which that rate might be altered over time. In general, the tax rate should be set with three factors in mind: (a) the rate should not be so high as to reduce investors' earnings to the point where it retards deep seabed mineral development; (b) the rate should not be set so as to significantly distort production between land-based and deep seabed mineral sources; and (c) the rate should be set so as to collect the rent element of production for international redistribution to developing nations. In terms of taxation rate flexibility, either one can argue for fixed rates which would then be subject to possible changes by the ISRA over time, or one can propose some form of built-in flexibility.

To examine the issue of the harmonization of land and seabed taxes briefly, it is argued that to fail to do so allows resources to be exploited inefficiently, since a lower tax–higher cost deposit could be exploited prior to a higher tax–lower cost deposit [63]. To minimize such potential economic inefficiencies in mineral production, it has been argued that seabed taxation levels should not exceed those taxes on land-based mineral sources. It is clear that while some distortions may develop, existing taxation rates on land-based raw materials are not equalized. In the case of copper, for instance, diverse national rates of taxation may distort production among alternative sources of supply. Furthermore, if national rates represent countries' particular preferences between present and future consumption, those taxations rates will inherently be different [64]. The problem is compounded since nodules contain a number of different minerals, so that any attempt to equalize land and seabed taxes would involve some sort of weighted average calculation for seabed taxes. What is likelier in the long run is that, with greater flows of market information and greater competition in resource markets, tax harmonization may come about naturally.

The issue of setting deep seabed mining taxation at levels high enough to generate revenue for developing nations yet not so high as to adversely affect current production of (or future investment in) deep seabed minerals can be joined with the taxation rate flexibility issue fairly easily. After all, as economic circumstances change, so must the tax level if producers and consumers are both to gain. Garnaut and Ross (G–R) assert that the prime deterrent to effective taxation policy is uncertainty [65]. G–R argue that the resource owner seeks not only to extract the maximum economic rent from the producer, but also seeks to attract new investments; thus a tradeoff exists between maximum rent extraction on current projects, which may involve the discouragement of future investment, and lesser rent extraction, but a larger number of projects to which investors are attracted. Were there perfect knowledge about the maximum permissible rent consistent with new investment in minerals development, there would be no dilemma. Unfortunately, however, uncertainties exist and knowledge is imperfect, especially during the first few years of production. G–R state the following proposition:

> Hence, in real world situations, where there is a higher degree of uncertainty about pre-tax profitability based on ignorance of costs as well as prices, a price-based system is inadequate, and a high proportion of the potential rents cannot generally be collected by governments (*resource owners*) except through a system of taxation that is based on both costs of production and prices, considered ex post [66].

Their solution is a proposed "resource rent tax," defined as a profits tax that begins to be collected when a particular threshold internal rate of return has been achieved. In principle, if the threshold were 25 percent, then each year the firm would be entitled to that return, and the excess—or preferably a portion thereof to encourage profit maximizing behavior—would be taxed. Profits would not be taxed twice. The tax could also be graduated, say x percent would be taxed between a 25 percent and 50 percent return and then $(x+y)$ percent would be taxed on the return exceeding 50 percent [67].

A slightly similar scheme is proposed by Charney to stimulate production through use of "sliding royalty" [68]. Such a proposal seeks to encourage production as the costs of extraction begin to rise through a reduced level of taxation. Presumably the same argument would apply at the initial stages of production when costs are also higher. In a sense, a fully sliding royalty system approaches a profits tax or a resource rent tax in the extreme case of frequent adjustments to reflect cost changes; however, it works through the revenue side.

A trade-off exists between the degree of flexibility in the establishment of a tax and the rate of taxation. If miners wish a riskless investment climate with a fixed contract which may be set for decades at a time, then the G–R type of scheme makes sense from the viewpoint of the Authority (if the method of calculating profits can be standardized). If, however, firms are not willing to give up most of the economic rent and resist such a return guarantee, then contracts with greater flexibility on the part of the resource owner to adjust the tax (to reflect the variability in the economic rent over the life of the investment) are likely in any multilateral taxation system.

The administration of an international deep seabed mining tax, however devised, can be a cumbersome operation. As previously indicated, certain types of taxes are easier to implement than others; still, a central apparatus would be required or the international community would need to rely on national governments to monitor firms and collect taxes in accordance with internationally agreed to and supervised rules. In the absence of standardized accounting procedures and the like, it would probably be practical and less costly to charge sponsoring states with the responsibility to collect revenue for the Authority [69].

There are a number of interesting points which arise out of such a direct involvement of states in the tax collecting process. Most of these are of national rather than international interest. For example, what kind of national tax treatment do tax payments to the Authority receive? Obviously some type of credit must be given or the firm will be burdened with dual taxation. Second, do states merely collect the international tax obligation and transfer the funds (in agreed to convertible currency) to ISRA or the ultimate dispenser, or do they deduct the funds from normal tax obligations (viz., this is especially relevant for the royalty-cum-profits tax). In a number of countries, there may be legislative barriers to the payment of international tax obligations which result from being signatory to a Seabed Treaty [70]. There are other considerations, as well, in how these taxes are treated; for example, national treasuries are more likely to view these tax payments as a form of development assistance if they pass through national budgets, while if they were paid directly (as are taxes by foreign investors), they would not be so viewed. A final concern is the "flags of convenience" problem; namely, deep seabed miners may prefer a particular sponsoring state because of lower tax rates or preferable treatment of the international seabed tax, although this then depends on the tax reciprocity relationship between the state of origin and the state of convenience.

INTERNATIONAL DEEP SEABED TAX REVENUES

This section approaches the issue of seabed tax revenues from both the supply and demand sides—that is, from the viewpoint of possible demands upon seabed revenue and projections of those possible revenues. There are a number of possible uses for the tax revenues which might be collected from deep seabed mining, among them are the operating expenses of ISRA, the diversification assistance contingency fund previously outlined in Chapter 6 and general revenue sharing or international sharing in the gains of deep seabed mining. The cost of operating the Authority depends crucially on the power to be vested in the Authority; however, at even a modest level of operations, ISRA would require a technical staff as well as support equipment. It is difficult to anticipate what costs might be; however, by way of comparison, the 1972 annual operating budgets for other United Nations specialized agencies were as follows: (1) International Civil Aviation Organization—$11 million; (2) International Labor Organization—$41 million; (3) World Meteorological Organization—$5 million; (4) International Atomic Energy Agency—$17 million; and (5) Food and Agriculture Organization—$47 million [71]. It is not clear to what extent general United Nations funds might be available to support ISRA, but it is not unreasonable to assume that $20 million in 1974 dollars might be required annually by 1985 to operate the Authority.

Estimates of funding levels for the diversification assistance contingency fund are also quite speculative. Much depends on the nature of the eligibility requirements and on the efficiency of existing institutions such as the International Bank for Reconstruction and Development in providing assistance for diversification type projects and the International Monetary Fund in providing compensatory financing funds. It is clear that there is a need for anticipatory diversification assistance for Zaire, which was projected to lose $82 million in 1985 in reduced cobalt earnings, and other cobalt producers. It is not expected that deep seabed taxation would generate sufficient funds to compensate major mineral exporting LDCs for actual losses, nor is that the purpose of the Fund. Therefore, budgeting for the 1981–85 period should include a major export diversification project in Zaire, which we provisionally list at $50 million, a reserve of $150 million for similar possible efforts elsewhere in the 1981–1985 period, and $50 million to subsidize 1981–85 IMF compensatory financing borrowing by developing countries whose mineral export earnings unexpectedly fluctuate in the short run because of nodule mining [72]. Therefore, a very tentative and crude guess at a budget for diversification assistance would require perhaps $50 million per year starting in 1981.

In forecasting possible revenue sharing potential in 1985, it should be borne in mind, therefore, that a mark of perhaps $70 billion must be made against any gross level of international taxation proceeds for these aforementioned purposes. In Table 7–1 gross revenue is calculated (under our usual assumptions) for various levels of profits tax and a royalty on production. Also listed are the amounts remaining for general revenue sharing after deductions for ISRA operating expenses and for the diversification fund. Last, the net post-international mining tax returns to equity are reported. Estimates for 1985 gross tax returns are between $99.7 and $177.9 million per 10 percent of profits taxed and between $165.6 million and $195.7 million per 10 percent of gross seabed mining revenue taxed. The total revenue sharing funds available for distribution would range from a low of $29.7 million with a 10 percent profits tax to a high of $641.6 million with a 40 percent profits tax; the corresponding after-tax returns to investment range between 79.2 percent and 29.6 percent. The total fund available from a royalty of 10 percent would be $65.6 million, while a 40 percent royalty on production would yield $832.8 million, calculated at final sales prices; the returns on investment, after-tax, would range between 76.9 percent and 16.6 percent.

By way of comparison, the United Nations estimates that a 50 percent profits tax would result in net returns in capital of between 22 percent and 55 percent for seven metal-recovery operations and between 25 percent and 47 percent for three-metal recovery operations [73]. Considering a 50 percent profits tax in our calculations, would yield a total return on all nodule investments of between 25 percent and 44 percent. The United Nations estimates for a 30 percent royalty yield post-tax profits of between 18 percent and 63 percent for seven-metal recovery operations and between 28 percent and 58 percent for three-metal recovery operations, compared to our return estimates of between 25 percent and 55 percent.

In attempting to gauge taxation levels which would be compatible with sustained investment in deep seabed mining, one may refer to the previously presented statistic that the 1974–75 average pre-tax return on mining investments in the United States was 27 percent [74]. Such investment returns are sustainable with profit taxes up to 40 percent in our example or with a royalty up to 30 percent, using the lowest revenue assumptions. Even including a risk premium for investments in deep seabed mining, it appears that a substantial level

Table 7–1. 1985 International Deep Seabed Taxation*

	Gross Tax Returns			General Revenue Sharing Funds			Average Net Return on Investment		
	Low	*Med.*	*High*	*Low*	*Med.*	*High*	*Low*	*Med.*	*High*
Profits Tax									
10%	99.7	138.1	177.9	29.7	68.1	107.9	44.4%	61.5%	79.2%
20%	199.4	276.2	355.8	129.4	206.2	285.8	39.4%	54.6%	70.4%
30%	299.1	414.3	532.7	229.1	344.3	463.7	34.5%	47.8%	61.6%
40%	398.8	552.4	711.6	328.8	482.4	641.6	29.6%	41.0%	52.8%
Royalty									
10%	165.6	195.7	225.7	65.6	125.7	150.7	41.1%	58.6%	76.9%
20%	331.2	391.4	451.4	461.2	321.4	381.4	32.9%	48.9%	65.6%
30%	496.8	587.1	677.1	426.8	517.1	607.1	24.7%	39.2%	54.5%
40%	662.4	782.8	902.8	592.4	712.8	832.8	16.6%	30.0%	43.4%

*Assuming fifteen million tons of nodules are mined and all projects except a single million-ton operation extract only three metals; assuming identical cost functions for all three metal-recovery operations as discussed in Chapter 6 and assuming metal prices and returns as reported on Tables 6–3, 6–4 (pp. 156, 160).

of international taxation could be absorbed by deep seabed miners, if national governments treated those taxes as either tax deductible cost items or reductions in calculating the national tax obligations assessed against miners [75].

The exact level of taxation will no doubt be set in a process of negotiation—negotiation over the division of economic rent—if a multilateral framework is to be forthcoming for the deep oceans. The level of taxation may be less a function of economic variables, such as the size of the economic rent or the sustainable taxation of miners or tax harmony among alternative mineral sources, than of political factors such as the strength of the developing country mineral producers bloc or the interrelationship of deep seabed issues with other Law of the Sea issues. Still, one may hope, if one believes in a negotiated multilateral regime for the deep seabed, that efficient mineral production can be stimulated with seabed producers, metals consumers and developing nations all sharing in the common heritage of mankind.

It has been pointed out by a number of observers of LOS issues that the level of revenue-sharing funds likely to be generated from the international taxation of offshore hydrocarbon exploitation could be expected to far exceed the proceeds of an international deep seabed mining tax [76]. It has also been forecast that the proceeds of deep seabed taxation will fall short of the net losses to developing country land-based mineral producers accruing as a result of deep seabed mining [77]. What might be added is that it is also very possible that both international offshore hydrocarbon taxes and manganese nodule taxes may be zero, if coastal states succeed in appropriating hydrocarbon resources seaward to 200 miles or more and industrialized countries unilaterally commence deep ocean mining. Certainly the potential magnitude is larger, if there is revenue sharing, in the case of hydrocarbons than in the case of manganese nodules [78]. Still, there are a number of important decisions which need to be made before any precise estimates are possible on the magnitude of revenue sharing funds which might be available from ocean resources.

The issue at present is will the developing nations, and especially the geographically disadvantaged nations, gain anything from ocean resources which are currently outside the limits of national jurisdiction. In Chapter 4 we discussed methods whereby the Third World might share in benefits to be derived from ocean resources [79]. Whether through direct allocation formulae or through development finance institutions, those concerned with development objectives would hope that benefits from ocean resource exploitation can be additional sources of resource transfers to developing nations [80]. In a world in which 50 percent of the world's population holds 10 percent of the world's production output, the international community now has the opportunity to ensure that ocean resources, considered to be the common heritage of mankind, are used in a manner which allows all countries to share in the benefits.

NOTES TO CHAPTER 7

1. On the regime, see E. Luard, *The Control of the Sea-bed*, (New York: Taplinger, 1975), Chapter 10–12 and J. Jacobson, "The United States and the Law of the Sea Conference: Interim Resource Management Options," in *Policy Issues in Ocean Law*, supra. pp. 121–184.

2. See pp. xviii for a discussion of the origin of the common heritage concept.

3. "Laissez faire" refers to the principle of government non-interference in the operations of economic markets.

4. See note 2.

5. John Mero, "The Great Nodule Controversy," in Christy, Clingan, Gamble, Knight and Miles (eds.), *Law of the Sea: Caracas and Beyond*, (Ballinger, 1975) p. 347.

6. *Ibid.*

7. See Group of Seventy-Seven Position Paper A/Conf. 62/C1/L7 (August 16, 1974).

8. For example, the U.S., which did not oppose the common heritage concept in the unanimously accepted United Nations General Assembly Resolution of December 17, 1970, is now reported not to accept the view that deep seabed resources are the common property of mankind. See Osgood, Hollick, Pearson and Orr, *Towards a National Ocean Policy* (National Science Foundation, 1975) p. 177.

9. In other worlds, optimal production occurs where marginal cost equals price.

10. The argument is roughly speaking that lower cost mineral inputs will, given perfect competition, lower the prices of final goods to international consumers including developing countries. This line of reasoning is based on the assumption that developing country consumers face perfectly competitive import markets, which in a world of large, highly integrated conglomerates is rather inaccurate.

11. A/Conf. 62/C1/L7 (August 16, 1974).

12. *Ibid.*

13. *Ibid.*

14. A/Conf. 62/C1/L6 (August 16, 1974). This proposal is similar to those submitted by the European Economic Community (excluding Ireland) and Japan, A/Conf. 62/C1/L8 (August 16, 1974) and A/Conf. 62/C1/L9 (August 19, 1974), respectively.

15. Resources are efficiently exploited if, other factors held constant, the lowest cost producer gains access to the resources and if, similar caveat applying, deposits are exploited in increasing order of costs. See Solow *op. cit.*

16. These proposals are often combined with work requirement provisions (e.g. minimum production levels) to insure against pure speculation by miners who merely acquire access to deposits in anticipation of future price increases. See UN, A/Conf. 62/C1/L6 (August 16, 1974).

17. Abnormal profits are those profits over and above those earned by the entrepreneur for his labors which would be necessary to keep the firm in business. In competitive market equilibrium, no single producer should be earning abnormal profits in the long run.

18. It should be noted that production efficiency is necessary but not sufficient to assure an optimal economic outcome, where that is defined in the Pareto-optimal sense. There are a number of conditions placed on producer and consumer actions to ensure a Pareto-optimal outcome (defined to be an allocation of resources which cannot be rearranged to increase anyone's welfare without decreasing someone else's welfare). Thus to discuss production efficiency in the absence of distributional efficiency issues is incomplete and at times midleading. For a general review of welfare economics, see Henderson and Quandt, *op. cit.*, Ch. 7.

19. All future references to the G–77 position stem from U.N. A/Conf. 62/C1/L7 (August 16, 1974).

20. The proposal includes, among others, the following provisions: (1) the ISRA will control all phases of deep seabed exploration and exploitation; (2) the ISRA may directly enter into any stage of deep seabed mining (perhaps through joint ventures, if it maintains majority ownership and administrative control) or enter into contracts for any mining activity; (3) the ISRA retains the power to regulate production levels; and (4) the ISRA would "obligate any mining contractor to "transfer to the Authority, on a continuous basis, technology, know-how, and data" relevant to the operations and to "undertake to provide at all levels training for personnel from developing countries, particularly the landlocked among them, and employment, to the maximum extent possible, to qualified personnel from such countries."

21. Evensen's remarks at the Caracas session of Committee One meetings referred to the United States proposal to allocate large areas of the deep seabed to miners for long periods of time (e.g., up to 55 years), while restricting the powers of resource managers (e.g., ISRA), to alter the rules and regulations of exploitation.

22. The proposal contains the following provisions, among others: (1) unlimited, first-come, first-served, access to deep seabed mining tracts for three stages of mining—prospecting (for which the ISRA would issue, automatically and in a non-discriminatory fashion, exclusive licenses to firms), evaluation (to be completed within 15 years), and exploitation (in a period of 20 years, renewable for an additional 20 years); (2) security of tenure for the investment if all conditions of the contract between the miner and ISRA are complied with; (3) freedom of independent operations, e.g., to set output levels; (4) work requirements in the form of mandatory annual expenditure to counter speculation in

mining sites; and (5) minimal supervisory responsibilities and access to non-proprietary information on the part of the ISRA. See U.N. A/Conf. 62/C1/L6 (August 16, 1974).

23. In particular, deep seabed mining firms are concerned with the positions of the developing nations on security of tenure, financial obligations, production controls, and the separability of mining activities. To be specific, the G–77 proposal gives ISRA the power to revise, suspend or terminate valid contracts in the case of a radical change in circumstances (i.e., "force majeure"); the right to control output; and the right to grant licenses for one stage of mining—say exploration—without guaranteeing a license for a subsequent stage of mining—say exploitation. This last provision would render it unwise for firms to invest in deep seabed mining ventures. In addition, the G–77 proposal includes scientific research and the processing, transportation and marketing of minerals within the ISRA's purview. In all of these areas, deep seabed miners could incur sizeable losses.

24. Although the exact makeup of an ISRA is unknown, it is logical to assume that the Group of Seventy-Seven would demand a major role in any such organ, as would the industrialized (including socialist) states. The question is who forms the majority and controls ISRA.

25. Economic rent may be defined as that return to a factor of production in excess of the amount necessary to keep the factor in the activity. In the case of firms, the rent accruing to the entrepreneur may be viewed as abnormal profits.

26. See Osgood, Hollick, Pearson and Orr, *op. cit.*, pp. 29–38, for a critical review of the negotiations.

27. This is based on the fact that only a few industrialized states are likely to possess the technology required for deep seabed mining for some time in the future. It should be noted that there is a good possibility that if there were more possible entrants into nodule mining, some industrialized nations might not prefer unlimited free access to deposits.

28. More general developing country objectives, such as those related to negotiating a "New International Economic Order," the fundamental restructuring of the world's economic system, are not examined.

29. For this view, see, for example, R. Eckert, "Exploitation of Deep Ocean Minerals: Regulatory Mechanisms and United States Policy," and R. Sweeney, R. Tollinson and T. Willett, "Market Failure, The Common-Pool Problem, and Ocean Resource Exploitation," *Journal of Law and Economics*, Vol. 17, No. 2 (April 1974), pp. 143–178 and pp. 179–192. See also F. Christy, "Alternative Regulation for Marine Resources Underlying The High Seas," *Natural Resource Lawyer*, Vol. 10, No. 2 (June 1968) and H. Demsetz, "Towards a Theory of Property Rights," *American Economic Review*, Vol. 57, No. 2 (May 1967), pp. 347–359.

30. See note 4, p. 142 for a discussion of prime sites.

31. See, for example, a statement by INCO, "An Approach to International Regulation of the Recovery of Deep Sea Ferro-Manganese Nodules," (January 1973).

32. Such legislation is currently before the United States Senate in the form of the Metcalf Bill (S. 713–1975) which would nationally regulate deep seabed

mining by United States firms. Reciprocal treaties might enable mining states to harmonize their nationally enacted regulations.

33. We do not share Pontecorvo's optimism in declaring that "... the development of the industry (manganese nodules) will benefit all mankind," since we are not convinced that market entry will be free enough to insure long-run competition in nodule mining. See G. Pontecorvo, "Reflections on the Economics of the Common Heritage of Mankind," *Ocean Development and International Law Journal*, Vol. 2, No. 3 (1974), p. 215.

34. United Nations Declaration of Principles, *supra.*, p. xxviiii.

35. See speech by Secretary Kissinger to the American Bar Association, "International Law, World Order, and Human Progress," (Montreal, August 11, 1975).

36. Certain Departments in the U.S. government are said to espouse this view.

37. Imagine, for instance, that a decade or two from now, a new type of deep seabed mining equipment is developed which involves significant pollution of the oceans.

38. The authority could be a profit maximizer or it could act to maintain the earnings of LDC land-based producers or to use deep seabed mining as a method of redressing the income disparity among nations. Much would depend on the voting mechanism to be established—would it be according to political interests of nations (e.g., one country, one vote as in the United Nations General Assembly) or according to economic interests (e.g., similar to the weighted voting in the I.M.F.). The 1970 United States Draft Treaty called for the establishment of an Assembly of all contracting parties; a Council of the six industrialized mining nations with highest per capita income and twelve LDCs among the twenty-four members which would include at least two land- or shelf-locked countries; a Tribunal of judges; and technical commissions of experts. Majority approval in the Council would require DC and LDC majorities taken separately as well. See "Draft United Nations Convention on the International Seabed Area," Working Paper by the United States (August 3, 1970).

39. See O.A. Roels, "The Environmental Impact of Manganese Nodule Mining," *Marine Technology Society Journal*, Vol. 8, No. 8, (September, 1974) and W. Garling and D. Leipziger, "Pollution in the Marine Environment: Interests of Developing Countries," mimeo, (April 1975).

40. To argue otherwise would be contrary to the common heritage principle; however, given the fact that LDCs are unlikely to be major metals consumers or direct participants in mining, one could argue that the DCs should voluntarily forego their share of such internationally distributed gains.

41. Hughes defines the resource rent of a deposit as "the difference between the cost of production for a given deposit and the cost of production of a marginal deposit, assuming all inputs and factor payments are priced so that their marginal cost equals marginal revenue." H. Hughes, "Economic Rents, The Distribution of Gains from Mineral Exploitation, and Mineral Development Policy," *World Development*, Vol. 3, No. 11–12 (1975), p. 812.

42. See p. 25, for a definition of economic rent. Also see Hughes, *op. cit.* for a discussion of monopolistic or quasi-monopolistic rents to factors of production and methods of appropriating those rents in mineral exploitation.

43. D. Logue and R.J. Sweeney, "An Economic Analysis of the Law of the Sea Negotiations," Department of the Treasury Discussion Paper, p. 13.

44. One method of altering contracts is through planned disinvestment, for example. Similarly, the international community might prefer to pick up equity positions in deep seabed mining ventures over time, in a type of joint venture, as a method of sharing in the gains of deep seabed mining.

45. R. Garnaut and A.C. Ross, "Uncertainty, Risk Aversion, and the Taxing of Natural Resource Projects," *Economic Journal*, Vol. 85, No. 338 (June 1975), pp. 272–287. See pp. 96–97.

46. See C. Allan, *The Theory of Taxation*, (Penguin, 1971) for a general review.

47. See H. Hotelling, *op. cit.* and R. Solow, *op. cit.* for discussions of optimal resource exploitation.

48. See the Trilateral Commission, *A New Regime for the Oceans*, 1976, p. 46.

49. Under a number of highly restrictive theoretical assumptions, one might expect increased raw materials supplies to benefit LDC consumers through normal trade and production. See R. Caves, *Trade and Economic Structure*, (Cambridge: Harvard University Press, 1960) for a discussion of international trade theories. For a review of studies on market imperfections see L. Weiss, "Quantitative Studies in Industrial Organization," in M. Intrilligator, (ed.), *Frontiers of Quantitative Economics* (New York: North Holland, 1971).

50. An annual level of mining of 15 million tons in the Pacific in 1985 would constitute approximately .9 percent of the estimated Pacific stock of nodules. At that rate of mining, the stock would last over a century, abstracting from the rate of accretion of the stock. See p. 124.

51. On the issue of revenue generation, see Charney, *op. cit.*, pp. 76–91 and J. Johnston, "Revenue Sharing With Respect to Minerals Recovered From the Seabed," Department of the Treasury, Mimeo (December 5, 1974).

52. See p. 188.

53. See Tobin, *op. cit.* and Garnaut and Ross, *op. cit.*

54. These costs discriminate against smaller firms and poorer nations.

55. Lump-sum taxes also raise revenue without distorting output and price variables. In economic theory, only taxes which affect the marginal costs of production (i.e., the increase in cost associated with a unit increase in output) affect the output decisions of firms and the short-run price of output in the competitive industry. If pre-tax output is said to be Pareto-optimal, then the tax is a distortion if it alters output. Lump-sum taxes do not alter output in the short-run since they are additions to fixed cost; royalties do affect output decisions; profits taxes—which apply to residuals of the production process—are not distortive in this sense.

While the distortion argument is sometimes brought into discussions concerning an appropriate deep seabed mining tax, it is a bit of a "red herring" in the sense that Pareto-optimal conditions are unlikely to prevail, especially in nodule mining, which will be controlled by a half dozen firms in the first years of mining. In the absence of Pareto equilibrium, a tax cannot be said necessarily to worsen the economic outcome. Lancaster and Lipsey state: "It is well known that the attainment of a Paretian optimum requires the simultaneous fulfillment of all the optimum conditions. The general theorem for the second best optimum states that if there is introduced into a general equilibrium system a constraint which prevents the attainment of one of the Paretian conditions, the other Paretian conditions, although still attainable, are, in general, no longer desirable. In other words, given that one of the Paretian optimum conditions cannot be fulfilled, then an optimum situation can be achieved only by separting from all the other Paretian conditions. The optimum situation finally attained may be termed a second best optimum because it is achieved subject to a constraint which by definition, prevents the attainment of a Paretian optimum." See K. Lancaster and R. Lipsey, "The General Theory of Second Best," *Review of Economic Studies*, Vol. 24, No. 63, (1956–57), pp. 11–32.

56. Some would argue that this is as it should be; why should the international community share in the value added to raw nodules, which is included in a taxation of the metals contained in the nodules. The answer must be that society cannot accurately gauge the resource rent of nodules but deserves a share of the economic rent of nodule production, if it is to sanction a system with an inherent bias towards a few industrialized countries.

57. In taxing profits, one must be aware of the fact that factors of production can be earning economic rent, especially in highly integrated firms, and thus that residual profits may be smaller than one might expect.

58. See A. Nove, *The Soviet Economy*, (New York: Praeger, 1961) pp. 30–39, 135–41 for a discussion of profit in socialist economies.

59. See Charney, *op. cit.*, Johnston, *op. cit.* and note 43.

60. The more price inelastic the demand for a product the greater the proportion of the tax that can be passed on to consumers visa price increases. In the extreme cases, a totally price inelastic (elastic) demand relation will allow all (none) of the tax cost to be passed on to consumers. See also pp. 102–104 on tax incidence.

61. Certain developed countries which are non-participants in nodule mining, yet which are metals consumers may object to relinquishing a direct share in the gains if they end up funding the redistributive scheme. See note 40.

62. See Johnston, *op. cit.*

63. Economic theory dictates that lowest cost sources of supply be exploited first. See Solow, *op. cit.*

64. In a world of more unified time preferences, one might argue for an international minerals bank which would allow production to increase at a socially optimal rate to maximize world welfare. We are far removed from such a Utopian state of affairs, however.

65. Garnaut and Ross, *op. cit.*

66. *Ibid.*, p. 277.

67. *Ibid.* p. 278.

68. Charney, *op. cit.*, pp. 84–86.

69. This also gets around the possible legal dilemma of ISRA suing a firm for non-payment and allows the state to make the necessary modifications in national tax payments to eliminate distortions in net taxes paid between land-based and deep seabed minerals producers.

70. In the United States, for example, the House of Representatives authorizes appropriations, while the Senate ratifies treaties. On these issues, see Charney, *op. cit.*, pp. 111–118.

71. United States House of Representatives, "United States Contributions to International Organizations," 93: Washington, D.C. (November, 1973).

72. In terms of a loan to Zaire, it would be preferable to begin the anticipatory diversification assistance prior to 1981, when the first seabed taxation revenue could become available. One possible approach would be for ISRA to "borrow" the funds from those generated from offshore hydrocarbon taxation, if such revenue comes on-stream prior to 1981. Certainly the $50 million figure is arbitrary. The additional pool of $150 million may or may not cover diversification assistance activities in the period depending on the criteria for eligibility and the actual level of LDC export earnings reversals. If the $50 million for subsidization of the IMF Compensatory Finance Facility were to be applied to ten-year loans (at 6.5 percent per annum), then the $50 million would be sufficient to subsidize all interest payments of $128 million in loans.

73. U.N. A/Conf. 62/25 (22 May 1974), *supra.*

74. See p. 159.

75. See p. 196 for discussion of this point.

76. See Charney, *op. cit.*, pp. 67–75.

77. See United Nations, "The Effects of Possible Exploitation of the Seabed on the Earnings of Developing Countries from Copper Exports," UNCTAD/TD/B/484 (28 May 1974), p. 9.

78. See p. 106.

79. See pp. 111–114.

80. Compared with the target level of Official Development Assistance to GNP of .7%, set by the Development Advisory Committee of the O.E.C.D., the average figure for DAC members in 1974 was only .3%. See, O.E.C.D., *Development Cooperation: 1975 Review* (Paris, Nov. 1975), p. 131.

Chapter 8

The Law of the Sea Negotiations: The Closing Phase

The preceding seven chapters have been devoted to a critical examination of the economic and technical aspects of two significant classes of subsea minerals in an attempt to explore how these minerals relate to the developing nations. At the same time that the research and writing processes were underway, sessions of the Third United Nations Law of the Sea Conference were also in progress. This meant that in order to pursue the examination it was necessary to make certain assumptions about key legal-institutional measures which could be expected to emerge from the Law of the Sea negotiations. The third session of the Third UN Law of the Sea Conference was held between March and May 1976 in New York, but in contrast to the initial anticipation of the conference planners, the third session was not the final session; it was decided to convene a fourth and presumably final session in New York to start some time in August 1976. The March–May New York conference began with the benefit of a *Single Negotiating Text*, which was the product of the preceding sessions and of extensive intergovernmental communication between formal sessions. This session failed to produce a Law of the Sea Treaty, but the third session yielded a *Revised Single Negotiating Text* (*Revised Text*). A review of this document indicates that the main assumptions about zones, revenue sharing, and the ISRA, presented in the previous seven chapters, remain essentially intact. Furthermore, the deliberate and broad-based development of the *Revised Text* suggests that its contents will comprise a major portion of the seabed treaty, if one is eventually produced. Thus, with the single exception of a brief digression concerning the synthesis of the two main parts of the book, the remainder of this chapter will focus on the efficiency and equity implications of provisions in the *Revised Text* which relate to the economics of seabed minerals and the economic interests of developing nations.

The examinations of hydrocarbons and manganese nodules have been presented in separate discussions, and there has been no attempt to combine them. The reason for this is straightforward: there is no significant economic connection between the two resources. Hydrocarbons and manganese nodules could conceivably be related in a political or institutional context, if there were a decision to make them related. A review of the *Revised Text* suggests, however, that an institutional relationship is not likely to be established, a "non-event" which also suggests that there have been no substantial political efforts to link the two resources. Thus, a synthesis of the two discussions at this time would not be particularly productive and, therefore, none is offered. Should a link between the two resources be developed, for example some kind of common revenue-sharing arrangement, then certainly an examination would be in order.

HYDROCARBONS

Throughout the examination of offshore hydrocarbon resources in the preceeding chapters, it was generally held that coastal states would in all likelihood acquire control over virtually all of these resources. While the precise nature of the measures necessary to implement coastal state control has not yet been finalized in the negotiations, it is becoming more certain that, as postulated, coastal states will acquire exclusive economic zones extending 200 nautical miles from base lines on shore. This issue is clearly delineated in the *Revised Text*, which uses two overlapping devices to establish coastal state control over hydrocarbon resources. The "Exclusive Economic Zone" is the first device. The economic zone would extend seaward 200 nautical miles from the coastal states' shoreline and, *inter alia*, the coastal state would have "sovereign rights for the purpose of exploring and exploiting, conserving and managing the natural resources, whether living or non-living, of the bed and the sub-soil in the super adjacent waters" [1].

The second device is a provision which gives coastal states "sovereign rights [over the continental shelf] for the purpose of exploring it and exploiting its natural resources" [2]. The continental shelf of a coastal state is defined as "sea-bed and sub-soil of the submarine areas that extend beyond its territorial sea throughout the natural prolongation of its land territory to the outer edge of the continental margin, or to a distance of 200 nautical miles . . ." [3].

Another basic issue in the discussion of hydrocarbons has been whether the Law of the Sea negotiations would produce a meaningful revenue-sharing measure. The revenue-sharing provision contained in the *Revised Text* indicates that revenue sharing is not likely to be a prominent feature of offshore hydrocarbon exploitation. Revenue sharing from hydrocarbon production is restricted to the area seaward of the exclusive economic zone (i.e., the area seaward of 200 nautical miles from shore) and sharing would occur only after the fifth year of production. The relevant section reads:

> The payments and contributions shall be made annually with respect to all production at a site after the first five years of production at that site. For the sixth year, the rate of payment or contribution shall be . . . per cent of the value or volume of production at the site. The rate shall increase by . . . per cent for each subsequent year until the tenth year and shall remain at . . . per cent thereafter. Production does not include resources used in connection with exploitation [4].

It may be recalled that a critique of this revenue-sharing proposal (see Chapter 4, pages 100–102) concluded that its revenue-generating potential was extremely small, and in view of the expected cost of production for most areas seaward of 200 nautical miles it is plausible that virtually no revenue would be generated. It may also be recalled that profitability tends to be highly sensitive to a gross production tax: i.e., a gross tax is a powerful deterent to production in a low-profit situation such as petroleum production beyond 200 nautical miles is likely to be.

With regard to efficiency implications, the economic zone and continental shelf provisions, devoid as they are of revenue-sharing measures, exert virtually no effect on the production efficiency of offshore operations. Because the economic zone and coastal shelf provisions essentially nationalize seabed hydrocarbon resources, there is no change in production efficiency from what it would be in the absence of an international agreement (assuming deep-water offshore exploitation would take place under national auspices). The application of a gross revenue sharing tax to crude hydrocarbon production seaward of 200 nautical miles could have an adverse effect on production efficiency in that area; however, the effect on aggregate deep-water offshore operations would be extremely small.

The implications for equity considerations of the zone, shelf and sharing provisions in the *Revised Text* are substantial. In Chapter 1 the pattern of offshore petroleum potential was examined and by means of correlation analysis compared to the existing world-wide pattern of income distribution. It was found that virtually all the poorest countries did not have any offshore petroleum potential (viz., most do not have coastal areas) and that a small but statistically significant positive correspondence exists between countries with low per capita income and countries with low (or zero) offshore petroleum potential. To the extent that the *Revised Text* confirms the premise that low-income and resource-poor economies without appreciable offshore petroleum potential will be excluded from direct or indirect participation in the receipt of gains, the final outcome of the Law of the Sea negotiations will exert a definite impact on the existing pattern of the worldwide income distribution.

It is possible to illustrate the general effect of the proposed exclusive allocation system on the distribution pattern by means of a Lorenz curve. In Figure 8–1, the horizontal axis measures the percent of world population and the vertical axis alternatively measures the cumulative proportion of income or offshore hydrocarbon potential. Curves *I* and *R* show graphically what proportion of in-

Figure 8–1.[a] Distribution of Offshore Hydrocarbon Resource Potential and Income Relative to Population

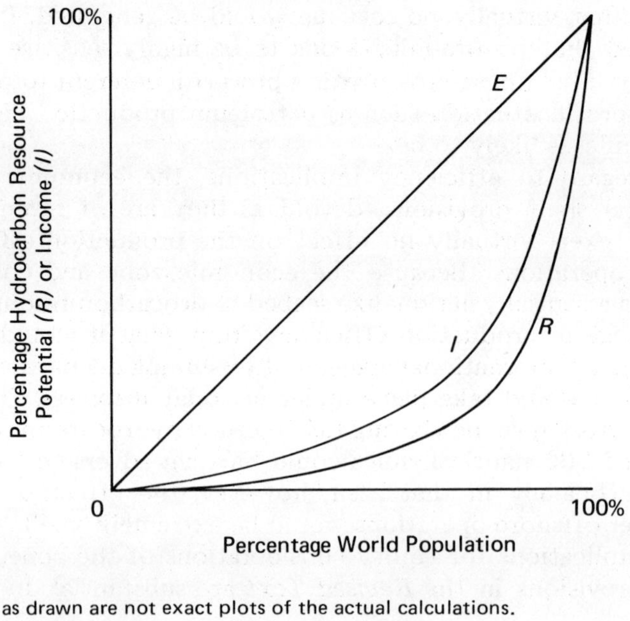

[a]The curves as drawn are not exact plots of the actual calculations.

come (I) and what portion of offshore petroleum potential (R) are controlled by a given portion of world population. Line E is drawn at an angle of 45 degrees with respect to the axes to represent a perfectly proportional distribution. The more the Lorenz curve bends away from line E, the greater the inequality of distribution. The Gini coefficients, arithmetic measures of the distribution equalities which are graphically depicted, are 0.648 for income relative to population and 0.840 for offshore petroleum potential relative to population [5]. The Gini coefficients indicate that the distribution of offshore petroleum potential is more unevenly distributed than the distribution of income is with respect to population.

The fact that income and offshore oil potential tend to be unevenly distributed with respect to population does not necessarily imply that the high-income population is also the resource-rich population. Indeed, it is conceivable that offshore petroleum potential could be concentrated off the shores of the poorer countries, thereby tending to offset partially the prevailing income pattern, i.e. *ceteris paribus* cause curve I to bend away from line E to a lesser degree and lower the Gini coefficient for income.

Whether curve R tends to further bend or to flatten, curve I can readily be determined by computing the sign and degree of correlation between per capita income and per capita offshore petroleum potential [6]. The correlation coefficients in Table 8–1 indicate that there is a small but positive relationship between per capita income and per capita offshore petroleum potential; and, the coefficients are statistically significant. This relationship applies to the entire 144-country group as well as to the two sub-groups which are split by per capita income. This positive correspondence means that if offshore oil wealth could be transformed into income, the global distribution of income would tend, *ceteris paribus*, to become more skewed in

Table 8–1. Comparison of Income per Capita with Petroleum Potential per Capita

	Correlation Coeff. (r)	*t-statistic*
All Countries (144)	0.424	5.58[a]
Countries with Per Capita Income > $1,200 (37)	0.353	2.42[b]
Countries with Per Capita Income < $1,200 (107)	0.443	4.92[a]

[a] significant at the 1 percent level.
[b] significant at the 2.5 percent level.

favor of the high-income countries (curve *I* would tend to bend further away from line *E*). Thus, the Law of the Sea Treaty, if adopted in the form expressed in the *Revised Text*, will have achieved the dubious distinction of allocating the largest portion of the single greatest stock of seabed mineral wealth to the relatively richer countries of the world and will thereby have contributed to a further widening of the gap between rich and poor nations.

MANGANESE NODULES

The Law of the Sea Conference has reached the stage where the developed countries are beginning to make concessions concerning the distribution of gains from deep-seabed mining and developing countries are similarly becoming more flexible and more concerned with their direct participation in manganese nodule mining. Thus the efficiency-cum-equity issues are finally being addressed in a more flexible negotiating atmosphere; however, there is some reason to believe that the resulting compromises will not necessarily result in solutions which are either efficient or equitable. We intend to explore the characteristics of the deep-seabed mining regime which appears to be emerging from the negotiations.

The major development during the Third Session of the Law of the Sea Conference, with respect to the deep-seabed negotiations, was a proposal introduced by the United States to break the impasse between developed and developing countries. The major aspects of the proposal are reflected in the *Revised Text*. Secretary of State Kissinger presented the new United States proposal, which included the following points: (1) acceptance by the United States of the "Enterprise" concept under which the ISRA can have an operational arm with the right to directly exploit the deep seabed under the same conditions as other miners; (2) reservation of mine sites for exploitation by the Enterprise or developing countries; (3) endorsement of the principle of revenue sharing, based either on royalties or profit sharing, with a portion of the revenues to be used for adjustment assistance through development finance institutions; (4) limitation for a number of years (perhaps two decades) of the level of total seabed production to respond to the concerns of land-based mineral producers, with the limitation to be tied to the growth rate of world nickel demand; and (5) reconfirmation of the 1970 U.S. proposal for an ISRA political structure to include a tribunal, council, assembly, and secretariat [7].

The concession of greatest import to the Third World, in its effort to obtain direct participation in deep-seabed mining, was the position of the United States to endorse mining by both private firms and the Enterprise, the operational arm of the Authority. The Enterprise concept has significant psychological and political meaning for developing nations as a method of rendering the "common heritage" principle operational.

In terms of efficiency, the Enterprise concept is troubling insofar as it would be controlled by the Authority and could potentially receive favorable discriminatory treatment vis-à-vis other miners [8]. If the Enterprise is the primary means whereby the developing nations share in the benefits of deep-seabed mining, then the Authority might come under pressures to act to ensure its success. As a new and experimental venture, the Enterprise might be subsidized (i.e., discriminatory treatment), perhaps using the "infant industry argument" as a justification [9]. If, through inexperience, conflicting objectives, market conditions, or other factors, the Enterprise flounders, the Authority may also act in a discriminatory fashion to protect it; this would be to the detriment of production efficiency, as private miners would receive no such benefits.

In terms of equity, it is not certain that direct participation is preferable to strong international taxation of deep-seabed mining. First, direct participation in mining requires initial capitalization—and assumption of risk—which could be achieved either through contribution, equity sale, or diversion of funds to the Enterprise from those generated by international taxation of deep-seabed mining. The sale of equity in the Enterprise would mean that the richest nations—industrialized, OPEC, or richer developing—could dominate the Enterprise and would receive most of its profits. Diversion of tax revenue from revenue-sharing purposes to the capitalization of the Enterprise involves a decision not to distribute revenue-sharing funds (viz., a significant portion of which would probably be directed toward the poorest nations), but instead to invest in a risky venture (the ownership pattern of which would probably favor richer nations). It is not clear that such a decision would not result in lower additional revenues for many developing nations, especially the poorest developing countries. If such a scenario develops, the equity concerns of the lower-income developing nations would not be adequately addressed within the Enterprise framework.

The bargaining power of the developing nations for a strong system of international taxation to capture economic rent is obviously reduced, if they opt, as they apparently have, for a chance at directly capturing a share of the benefits of deep-seabed mining. If the Enterprise can raise the necessary capital and can operate as efficiently as private industry, it may indeed garner a larger share of the benefits of mining than would international taxation; however, this is by no means certain. Quite apart from how distributional shares in the Enterprise's profits are allocated, the Third World is wagering—and it is quite a gamble—that it can extract a larger portion of the economic rent involved in nodule mining via sub-contracting or independent ventures than it could extract via direct taxation of private miners [10].

The second important development of the fourth session in New York was the endorsement of reservation of mine sites by the United States, a prominant spokesman for most of the industrialized economies. Reservation of mine sites involves the selection by the Authority of one of two possible sites desired by miners for use by the Enterprise at a later date. This arrangement is potentially inefficient if the holding of prime sites by the Enterprise forces other miners eventually to exploit less than prime areas while prime sites lie idle [11]. The parallel system is also potentially discriminatory inasmuch as the Enterprise is spared the costs of exploration; this is an example of one of the possible benefits accruing to the operational arm of the authority under a parallel system of exploitation [12].

It is interesting to note that there is a difference of opinion among the major industrialized nations over whether there should be limits on the number of mine sites claimed by firms from particular countries. In addition to reserving sites for the authority's use, some industrial country delegations wish to establish national quotas on mine sites, presumably to restrain the United States from establishing too dominant a position with respect to deep-seabed mining sites. The United States position is that such quotas are unnecessary inasmuch as there are a sufficient number of prime mine sites to prevent monopolization of nodule mining by any single country. Once limits on the number of operations by individual countries are imposed, however, rather than making sites available to the most efficient producer, a radically different system of exploitation evolves from that which the United States has traditionally advocated [13].

Point three of Secretary Kissinger's address included U.S. endorsement of the principle of international taxation, with a portion of the revenue earmarked for adjustment assistance. The particular method of international taxation proposed by the United States at the New York session is very similar to the previously discussed guaranteed rate of return method proposed by Ross and Garnaut [14]. The United States proposal includes a variable tax on net proceeds (defined as revenue minus operating costs and development costs, the latter being cost of construction and equipment) depending on whether the return on investment is below, at, or above the "normal" or threshold level [15]. Presumably, the rate of taxation would increase as the rate of return increases and therefore would be similar to a variable profits tax. The major difference, however, is that taxes are based on the average profitability of the venture over time, so that losing or lean years can be averaged with profitable years in calculating the return. Precisely what the taxation rate should be, how profits are to be calculated for firms from different countries, and how revenue should be allocated among objectives (i.e., the Enterprise, the ISRA and adjustment assistance) or recipients (i.e., developed and developing countries) must still be determined. Whether such detailed provisions can be negotiated in the fourth Law of the Sea session in New York is doubtful, but not impossible [16]. While it is impossible to prejudge the conference outcome, one possibility is that if agreement is reached on the fundamental structure of a deep-seabed mining regime, technical groups will be established to thrash out the very important details.

The fourth aspect of the United States proposal which dominated the third session was the advocacy by the United States of a limitation on the future level of seabed mining, supposedly in response to the concerns of the land-based mineral producers. The United States proposal, which appears in Article 9 of the *Revised Text*, would seek to limit seabed production so that it does not exceed the projected cumulative growth segment of the nickel market during the period in question (perhaps for twenty years), provided the growth rate is at least 6 percent per year [17]. In essence, the "limitation" would operate in the following fashion. Let us assume that world nickel production satisfies nickel demand and expands at a projected annual rate of 6 percent from 1975 to the year 2000; worldwide production in 1980 would then be 1048 thousand metric tons (tmt), all of which would be supplied by land-based sources. Seabed production would begin in 1980, and in the year 2000, seabed production could not exceed the total cumulative increase in nickel demand since 1980. In 1985, for example, the proposed production limit would be 335 tmt, a level of seabed nickel production which is equivalent to approximately twenty-four nodule operations which each yield one million tons annually and which is well below current projections of 1985 nodule mining levels [18]. By 1990, the limit would be 834 tmt or the equivalent of fifty-six one-million-ton nodule operations.

Whether this is an operative constraint or not is uncertain [19]. The basic concept of such a limitation is highly questionable, however. In particular, to base the limitation of joint-product mining on any one mineral's consumption is rather puzzling, since it ignores the effect of joint mineral production on other mineral markets. In addition, to select nickel as the constraint under the guise of protecting developing country land-based producers ignores the fact the LDCs are not in general major producers of nickel. The mineral which could most seriously affect particular developing country producers is cobalt, and the proposed 1985 limit of 355 tmt of nickel corresponds to a level of cobalt production of sixty tmt or 100 percent of projected 1985 demand. Thus seabed-produced cobalt would completely swamp land-based production. The effects of the proposed mining limit on copper production, which is of concern to a number of developing countries, is to allow a maximum of 284 tmt in 1985. This corresponds to 3 percent of projected 1985 copper demand. Therefore, the proposed limit would not significantly affect output. The limit is similarly ineffective in affecting manganese ore output. Furthermore, as long as manganese nodule mining is not projected to increase at an annual rate exceeding 6 percent per annum on a 1980 base of 1048 tmt—equivalent to an average increase of five to six million tons of nodules per year in the first decade—the constraint is not an operative one. It clearly does not adequately address the concerns of developing country mineral producers whose economies may be adversely affected by manganese nodule mining.

The final proposal offered by the United States concerns the structure of the International Seabed Resource Authority. A discussion of the organs of the authority—the Assembly, the Council, the Tribunal, special commissions, and the Secretariat—is contained in the *Revised Text.* The key to the acceptability of a treaty to individual states lies with the structure of these organs; the United States proposes that the council, the administrative organ, be controlled by producer and consumer interests, a proposal which would favor the industrialized nations. Given the powers of ISRA to license miners, control the Enterprise, set production limits, set international taxation levels and allocate tax funds, the entire negotiations can be expected to coalesce into the issue of who controls the Authority. And the Third World is unlikely to agree on a minority position vis-à-vis control of the Seabed Authority.

In summary, the deep-seabed negotiations are progressing toward greater involvement of the Third World in deep-seabed mining. In terms of implementing the principle of common heritage, this is encouraging; however, the proposed system of parallel exploitation poses possible infringements on production efficiency without guaranteeing developing countries any equity gains. If equity investment in the Enterprise determines the distribution of returns, then it is likely that the poorest nations will receive fewer benefits in the system which appears to be emerging than they would receive from a strong system of international taxation and revenue sharing. The central issue is whether those nations financing the Enterprise consider themselves donors or investors. Aspects of the *Revised Text* on the deep-seabed mining regime may be responsive to the desires of the Third World for direct participation in exploitation of manganese nodule resources; however, the end result may be a package labeled "deep-seabed resources for the majority of poor nations," but containing an essentially empty box.

SUMMARY AND CONCLUSIONS

The main points evolving from the analysis and the negotiations can be reiterated briefly. With respect to production efficiency for seabed hydrocarbons, it can reasonably be expected that production efficiency will not be less than it has been in the past. With respect to manganese nodule production efficiency, there is no past, but the *Revised Text* appears to contain a number of elements conducive to unencumbered mining. The extent to which mining will be efficient will depend on the precise structure and powers of ISRA and on the interrelationship between private enterprise and ISRA mining. The detractions to production efficiency introduced by direct Enterprise exploitation should be of a lesser magnitude than alternatives such as production controls, which were originally discussed. For a number of reasons, it is not expected that developing nations will play a major, direct role in either seabed hydrocarbon or manganese nodule recovery operations.

On the equity side, with respect to hydrocarbons, it appears that the distribution of benefits will follow the pattern in which these resources have been distributed by the forces of nature and this pattern roughly coincides with the distribution of other natural resource wealth and the distribution of national income among the countries of the world. If the position is taken that the physical distribution of hydrocarbon resources with respect to nation states represents the appropriate criterion, then there can be little complaint about what appears to be the Law of the Sea scheme for the international allocation of seabed hydrocarbons. It is noted, however, that there is no intrinsic basis for nationalization of these resources and that nationalization will effectively exclude approximately fifty-seven countries with nearly 9 percent of total world population. Of the fifty-seven countries, forty-eight countries have annual per capita incomes less than $500; they contain 7 percent of world population and they include almost all of the poorest economies of the world. Thus, there is a reason to criticize the pending distribution-by-nationalism scheme. Clearly, we have taken the latter stance on hydrocarbon issues and would conclude that the international allocation of this major stock of resource wealth is not international in any real sense, but represents an instance in which the national self-interests of most of the more powerful and wealthy countries of the world were sufficiently common to allow a majority consensus. The rights of the minority were not seriously considered and are not expected to be incorporated in the final results.

The equity case for manganese nodules is similar, although considerably more subtle, and the specific outcome is less certain. There are provisions to generate revenue for developing nations, but there are substantial reasons to doubt that the poor countries will receive more than token sums from manganese nodule exploitation. If the Enterprise is the major benefit distribution vehicle but distribution shares are determined by equity holdings in the Enterprise, then it would appear that ability to invest in the Enterprise would be the major determinant in directing the international distribution of benefits. It is too early to tell if the Enterprise will become "a rich man's club," but the possibility appears to be open. Certainly, the receipt by the Enterprise of revenue-sharing funds constitutes an intervening element which could work to deflect funds away from the poorest countries. Time and additional analysis will be required to tell if the interests of the poor countries are being transmitted through the Enterprise or if the Enterprise will be dominated by a select group. There is a chance the Enterprise will yield an even larger flow of benefits to low-income countries, but if their interests are not respected, then, as in the distribution of benefits from seabed hydrocarbon production, the poorer countries will find themselves watching both of the great mineral stocks of the oceans' floors being depleted by the relatively wealthy countries for the welfare of the wealthy countries.

NOTES TO CHAPTER 8

1. United Nations, *Third United Nations Conference on the Law of the Sea: Revised Single Negotiating Text*, New York Session, 15 March–7 May 1976, pt. 2, p. 26.

2. *Ibid.*, p. 36.

3. *Ibid.*, p. 35. The necessity for including the economic zone and continental shelf provisions, both of which are virtually identical as they pertain to subsea hydrocarbon resources, is not explained in the *Revised Text.* The inclusion of both provisions does, however, inspire some speculation. First, the presence of two provisions, both giving coastal states sovereign rights over sub-soil resources,

leaves little doubt about the significance attached by coastal states to acquisition of rights to resources in this area. Second, and in a more speculative mode, the inclusion of the continental shelf provision, while redundant relative to the economic zone provision, is unique in at least one interesting aspect. The definition of the continental shelf states that the shelf is a natural component of the adjacent coastal states; thus, the coastal shelf provision contains one of the few substantive notes in this formal claim on hydrocarbon resources located in a formerly international area. Reference to the continental shelf as "the natural prolongation of [the coastal states'] land territory. . . ." coincides precisely with Hedberg's original thesis. It also ignores the logical implications of the natural extension theme: national borders are not natural separations between countries; therefore, the continental shelf is also the natural extension of land-locked and geographically disadvantaged states' land territory as well.

4. *Ibid.*, p. 37.

5. The Gini coefficient is the ratio of the area between line *E* and the Lorenz curve divided by the entire area beneath line *E*. Perfectly equal distribution would yield a Gini coefficient of zero. For an explanation and critique of the Lorenz curve and the Gini ratio see Everett E. Hagen, *The Economics of Development* (Homewood, Ill.: Richard D. Irwin Inc., 1975), pp. 216–17.

6. A very similar exercise was carried out in Chapter 1 when the coefficients of correlation between national per capita income and national offshore petroleum potential were calculated. In that instance, national petroleum potential was used in the calculation of correlation coefficients to make the discussion of distribution more appropriate in an international political context. In this instance, per capita petroleum potential and per capita national income levels are used to correspond to the Lorenz-curve framework.

7. Address by The Honorable Henry A. Kissinger before the Foreign Policy Association, U.S. Council of the International Chamber of Commerce, and the United Nations Association of the United States of America (New York, April 8, 1976). Also, United States, "Draft United Nations Convention on the International Seabed Area," Working Paper (August 3, 1970).

8. See United Nations, *Third United Nations Conference on the Law of the Sea: Revised Single Negotiating Text, Annex II: Statute of the Enterprise*, New York Session, 15 March–7 May 1976.

9. This argument is often used for the subsidization of new, developing industries, with the rationale that the industry requires temporary assistance until it can establish itself and produce profitably.

10. Even if the clientele were the same and the relative shares in revenue distributed or in the Enterprise were identical, one might easily argue that poorer countries would prefer not to invest their share of revenue-sharing funds in risky deep-seabed mining ventures, but would rather favor direct allocations of funds for current consumption or alternative investment.

11. See Chapter 7, p. 183, on optimal exploitation.

12. Whether such cost reductions actually yield greater profits to be distributed to the Enterprise's share-holders is uncertain. It is indicative, however, of the kind of favorable treatment which accrues to the Enterprise.

13. While it is interesting to attempt to project the number of prime nodule sites available, such an activity is highly speculative and depends not only on scientific assumptions concerning the size, density, and content of worldwide nodule deposits, but also on the technical and economic factors which render a site a prime site. Among these latter factors are the cost of mining, the assumed size of sites and length and intensity of mining activities, the demand (prices) for particular minerals (especially nickel), and the availability of capital to undertake the mining project. Without denying that there are probably some prime sites which are more advantageous than others, it might be conservatively assumed that there are in excess of a hundred prime sites for first-generation-scale mining. See, for example, A.A. Archer "The Prospects for the Exploitation of Manganese Nodules: The Main Technical, Economic and Legal Problems" (Paper presented at the CCOP/SOPAC–IOC/IDOE International Scientific Workshop on the Geology, Mineral Resources and Geophysics of the South Pacific, Suva, Fuji, September 6, 1975).

14. See Chapter 7, pp. 196–198.

15. United Nations, *Third United Nations Conference on the Law of the Sea: Revised Single Negotiating Text*, Annex I: "Basic Conditions of Prospecting, Exploration and Exploitation," New York Session, 15 March–7 May 1976.

16. There remains the problem of what rules apply in the interim between any negotiated treaty and the date that such a treaty would come into force. The United States proposes that a provisional set of rules apply in the interim.

17. *Ibid.*, pp. 13–14.

18. The proposal would select a base year of 1980 and calculate the average growth rate of the previous three years, subject to a 6 percent minimum. At 6 percent annually, 1985 nickel demand should total 1403 tmt, of which 1048 is estimated to have been land-produced in the absence of seabed mining in 1980.

19. Certainly in the first decade of nodule mining (and perhaps the second), the volume of mining which can realistically be expected to be attained falls short of the proposed constraint.

Bibliography

Adams, F.G. *The Impact of Nickel Production From the Ocean Floor; An Econometric Appraisal.* United Nations Conference on Trade and Development, Geneva, November 20, 1974.

Adelman, M.A. "Is the Oil Shortage Real? Oil Companies as OPEC Tax-Collectors." *Foreign Policy* 9 (Winter 1972–73): 69–107.

——. "Politics, Economics, and World Oil." *American Economic Review: Papers and Proceedings of the Eighty-sixth Annual Meeting of the American Economic Association* 64 (May 1974):58–67.

——. "The World Oil Outlook." In *Natural Resources and International Development*, pp. 27–113. Edited by Marion Clawson. Baltimore: The Johns Hopkins Press for Resources for the Future, Inc., 1964.

Akins, J.E. "This Time the Wolf Is Here." *Foreign Affairs* 51 (April 1973): 462–90.

Alexander, Lewis M., ed. *The Law of the Sea: The United Nations and Ocean Management, Proceedings of the Fifth Annual Conference of the Law of the Sea Institute, June 15–June 19, 1970.* Kingston: The University of Rhode Island, January 1971.

Allan, C. *The Theory of Taxation.* Baltimore: Penguin Books Inc., 1971.

Alparone, Nicolò. "Gravity Platform Designed To Reduce Offshore Costs." *World Oil* 176 (February 15, 1973):90–97.

American Society of International Law. *Policy Issues in Ocean Law.* St. Paul: West Publishing Co., 1975.

Annual Statistical Report. Philadelphia: American Iron and Steel Institute, 1973.

Archer, A.A. "The Prospects for Exploitation of Manganese Nodules: The Main Technical, Economic, and Legal Problems." Paper presented at the CCOP/SOPAC–IOC/IDOE International Scientific Workshop on the Geology, Mineral Resources and Geophysics of the South Pacific Suva, Fiji, September 6, 1975.

Bergsten, C. Fred. "The New Era In World Commodity Markets." *Challenge* 17 (Sept.–Oct. 1974):34–42.

Blitzer, Charles; Meeraus, Alex; and Stoutjesdijk, Ardy. "A Dynamic Model of OPEC Trade and Production." *Journal of Development Economics* 2 (December 1975):319–35.

Boldt, Joseph R. *The Winning of Nickel: Its Geology, Mining and Extractive Metallurgy*. New York: Van Nostrand Reinhold Co., 1967.

Branco, R. "The Tax Revenue Potential of Manganese Nodules." *Ocean Development and International Law Journal* 1 (1973):201–208.

"Brazilian Oil Field." *Washington Post*, 6 December 1974, p. A9.

The British Petroleum Company Limited. *BP Statistical Review of the World Oil Industry, 1974: Reserves, Production, Consumption, Trade, Refining, Tankers, Energy*. London: Britannic House, 1975.

Caves, Richard. *Trade and Economic Structure*. Cambridge: Harvard University Press, 1960.

Charney, Jonathan I. "The Equitable Sharing of Revenues from Seabed Mining." In *Public Issues in Ocean Law: Studies of the Working Group on Technical Issues of the Law of the Sea*, pp. 53–120. St. Paul, Minn.: West Publishing Company for the American Society of International Law, 1975.

Chenery, H.B., and Strout, A.M. "Foreign Assistance and Economic Development." *American Economic Review* 56 (September 1966):679–733.

Christy, Francis T., Jr.; Clingan, Thomas K., Jr.; Gamble, John King, Jr.; Knight, H. Gary; and Miles, Edward, eds. *Law of the Sea: Caracas and Beyond*. Proceedings, Law of the Sea Institute, Ninth Annual Conference, January 6–9, 1975. Cambridge, Mass.: Ballinger Publishing Company, 1975.

Clauss, G. "Theoretical and Experimental Investigations of Deep Sea Ocean Mining Systems and Their Economic Evaluation." Paper presented at the Second International Ocean Development Conference, Tokyo, October 1972, pp. 1925–55. As reported in United Nations, "Economic Implications of Sea-bed Mineral Development in the International Area: Report of the Secretary General." A/Conf. 62/25 (22 May 1974), p. 67.

Cooper, Richard. "An Economist's View of the Oceans." In *Perspectives on Ocean Policy: Conference on Conflicts and Order in Ocean Relations, Arlie, Virginia, 21–24 October 1974*, pp. 143–66. Prepared for the National Science Foundation by The Johns Hopkins University Ocean Policy Project, Washington, D.C.

Cornell, Nina W. *Techno-Economic Focus Paper on Manganese Nodule Mining*. Washington, D.C.: Brookings Institute, April 1973.

Craven-Walker, A. "How BP Will Equip the North Sea's Forties Field." *World Oil* 176 (May 1973):55–60.

Crittenden, A. "Loans Abroad Stir Worry: U.S. Bank Write-Offs High." *The New York Times*, 15 January 1976, p. 1.

"Deep Ocean and Offshore Mining." *Mining Engineering* (April 1975).

Deep Sea Ventures Inc. "Notice of Discovery and Claim of Exclusive Mining Rights," and "Request For Diplomatic Protection and Protection of Investment." Gloucester Point, Virginia, November 15, 1974.

Delach, M.; Horn, B.; and Horn, D. "Ocean Manganese Nodules: Metal Values and Mining Sites." Technical Report No. 4, National Science Foundation, Washington, D.C., 1973.

Demsetz, Harold. "Towards A Theory of Property Rights." *American Economic Review* 57 (May 1967): 347–59.

Devanney, John W., III. "Key Issues in Offshore Oil. *Technology Review* 76 (January 1974):21–25.

Dougherty, E.L.; Goodknight, R.D.; and Oldaker, R.B. "A Comparative Study of the Economics of a Subsea and a Platform Oil Producing System." Paper presented at Oceanology International 72 conference in Brighton, England, 19–24 March 1972.

Dubs, M.A.; Flipse, J.E.; and Greenwald, R.J. "Pre-Production Manganese Nodule Mining Activities and Requirements." In U.S. Congress, Senate, Subcommittee on Minerals, Materials, and Fuels, Committee on Interior and Insular Affairs, *Mineral Resources of the Deep Seabed.* 93d. Cong., 1st sess. May–June 1973, pp. 602–709.

Eckert, R. "Exploitation of Deep Ocean Minerals: Regulatory Mechanisms and United States Policy." *Journal of Law and Economics* 17 (April 1974): 143–78.

Engineering and Mining Journal and Coal Age. Vol. 4. New York: McGraw-Hill Book Company, 1973.

Fenn, Percy T. *The Origin of the Right of Fishing in Territorial Waters.* Cambridge, Mass.: Harvard University Press, 1926.

Fischer, Dietrich; Gately, Dermot; and Kyle, John F. "The Prospects for OPEC: A Critical Survey of Models of the World Oil Market." *Journal of Development Economics* 2 (December 1975):363–86.

Flanders, M. June. *Toward Assessing the Need for International Reserves.* Essays in International Finance No. 58. Princeton: Princeton University Press, 1967.

Gamble, John King, Jr., and Pontecorvo, Giulio, eds. *Law of the Sea: The Emerging Regime of the Oceans.* Proceedings of the Eighth Annual Conference of the Law of the Sea Institute, June 1973. Cambridge, Mass.: Ballinger Publishing Company, 1975.

Garnaut, R., and Ross, A. Clunies. "Uncertainty, Risk Aversion and the Taxing of Natural Resource Projects." *The Economic Journal* 85 (June 1975): 272–87.

Gil, Federico G. *Latin American–United States Relations* New York: Harcourt Brace Jovanovich, Inc., 1971.

Glezakos, Constantine. "Export Instability and Economic Growth: A Statistical Verification." *Economic Development and Cultural Change* 21, pt. 1 (July 1973):670–78.

Griffen, K.B. "Foreign Capital, Domestic Savings and Economic Development." *Bulletin of the Oxford University Institute of Economics and Statistics* 32 (May 1970):355–69.

Hagen, Everett E. *The Economics of Development.* Homewood, Ill.: Richard D. Irwin Inc., 1975.

Hayward, Henry S. "Nigeria's Costly 'Flaring' of Gas. *Christian Science Monitor*, 24 October 1974.

Hedberg, Hollis D. "The National-International Jurisdictional Boundary on the Ocean Floor." *Ocean Management* 1 (1973):83–118.

——. "Why Explore the Deep Offshore?" In *Exploitation and Economics of the Petroleum Industry: New Ideas, New Methods, New Developments*, pp. 61–84. Houston: Gulf Publishing Company, 1967.

Hemphill, William L. "The Effect of Foreign Exchange Receipts in Imports of Less Developed Countries. *IMF Staff Papers* 21 (November 1974):637–77.

Henderson, James M., and Quandt, Richard E. *Microeconomic Theory: A Mathematical Approach.* Economics Handbook Series. Edited by Seymour E. Harris. New York: McGraw-Hill Book Company, Inc., 1958.

Hirschman, Albert O. *The Strategy of Economic Development.* New Haven and London: Yale University Press, Yale Paperbound, 1958.

Hotelling, Harold. "The Economics of Exhaustible Resources." *The Journal of Political Economy* 39 (April 1931):135–75.

Hughes, H. "Economic Rents, The Distribution of Gains From Mineral Exploitation and Mineral Development Policy." *World Development* 3 (Nov.–Dec. 1975):811–25.

"Increasing Demand for Oil Spurs Offshore Growth Rate. *World Oil* 177 (July 1973):82.

International Bank for Reconstruction and Development. *The Nickel Outlook Reassessed.* Economic Staff Working Paper No. 135, by B. Varon. Washington, D.C., August 31, 1972.

World Bank Atlas. Washington, D.C., 1975.

International Monetary Fund. *Articles of Agreement.* Washington, D.C., 1944, as amended by modifications adopted May 31, 1968. Reprinted 1973.

International Nickel Company of Canada, Ltd. *An Approach to International Regulation of the Recovery of Deep Sea Ferro Manganese Nodules.* Distributed Company Statement, New York, January, 1973.

International Petroleum Institute. *1985: A Dual Impact Study of Our World in 1985 before and after the Energy Crisis.* New York: M. Barrows & Co. and International Petroleum Institute, 1975.

Jacoby, Neil H. *Multinational Oil: A Study of Industrial Dynamics.* New York: Macmillan Publishing Co., Inc., 1974.

Kalymon, Basil A. "Economic Incentives in OPEC Oil Pricing Policy." *Journal of Development Economics* 2 (December 1975):337–62.

Kash, Don E.; White, Irvin L.; Bergey, Karl H.; Chartock, Michael A.; Devine, Michael D.; Leonard, R. Leon; Salomon, Stephan N.; and Young, Harold W. *Energy under the Oceans: A Technology Assessment of Outer Continental Shelf Oil and Gas Operations.* Norman: University of Oklahoma Press, 1963.

Kanenas. "Wide Limits and 'Equitable' Distribution of Seabed Resources." *Ocean Development and International Law Journal* 1, no. 2 (1973):137–58.

Kaufman, Raymond, and Arnold J. Rothstein. (Deep Sea Ventures, Inc.). "The Approaching Maturity of Deep Ocean Mining—The Pace Quickens." In U.S. Congress, Senate, Subcommittee on Minerals, Materials, and Fuels; Committee on Interior and Insular Affairs. *Mineral Resources of the Deep Seabed.* 93d. Cong., 1st sess. May–June 1973, pp. 201–22.

Kenen, P., and Voivodas, C. "Export Instability and Economic Growth." *Kyklos* 25, fasc. 4 (1972): 791–804.

Kennedy, Michael. "An Economic Model of the World Oil Market." *The Bell Journal of Economics and Management Science* 5 (Autumn 1974):540–77.

Kissinger, Henry A. Address before the Foreign Policy Association, U.S. Council of the International Chamber of Commerce, and the U.N. Association of the U.S.A., New York, April 8, 1976.

Knudsen, Odin, and Parnes, Andrew. *Trade Instability and Economic Development: An Empirical Study* Lexington, Mass.: Lexington Books, 1975.

Kyle, John F., and Moskowitz, Warren E. "The Economics of the OPEC Cartel: A Theoretical Discussion." New York University No. 74–20, October 1974. (Mimeographed.)

Labys, W.C., ed. *Quantitative Models of Commodity Markets.* Cambridge: Ballinger Publishing Company, 1975.

Lancaster, K., and Lipsey, R. "The Theory of Second Best." *Review of Economic Studies* 24, no. 63 (1956–57):11–32.

Le Leuch, H., and Masseron, J. "Economic Aspects of Offshore Hydrocarbon Exploration and Production." *Ocean Management* 1 (December 1973):287–325.

Luard, Evan. *The Control of the Sea-bed: A New International Issue.* New York: Taplinger Publishing Co., Inc., 1974.

"Who Gets What on the Seabed?" *Foreign Policy* 9 (Winter 1972–73):132–47.

"Major Oil Developments Programmed To Make Peru a Net Exporter within Two Years." *IMF Survey* 4 (28 April 1975):125–27.

Mamalakis, Markos. "Contribution of Copper to Chilean Economic Development, 1920–67: Profile of a Foreign-Owned Export Sector." In *Foreign Investment in the Petroleum and Mineral Industries: Case Studies of Investor-Host Country Relations.* Baltimore: Johns Hopkins University Press for Resources of the Future, 1971.

Massel, B. "Export Instability and Economic Structure." *American Economic Review* 58 (September 1970):618–30.

McBean, A. *Export Instability and Economic Growth.* Cambridge, Mass.: Harvard University Press, 1966.

McKelvey, V.E. "Environmental Protection in Offshore Operations." *Ocean Management* 1 (March 1973):119–27.

McKie, James W. "The Political Economy of World Petroleum." *American Economic Review: Papers and Proceedings of the Eighty-sixth Annual Meeting of the American Economic Association* 64 (May 1974):51–57.

McKinnon, Ronald I. "Foreign Exchange Constraints in Economic Development and Efficient Aid Allocation." *The Economic Journal* 74 (June 1974): 388–409.

——. "International Transfers and Non-Traded Commodities: The Adjustment Problem." In *The International Monetary System and Developing Nations.*

Edited by Danny M. Leipziger. Washington, D.C.: Agency for International Development, forthcoming.

Mero, John L. "The Great Nodule Controversy." In *Law of the Sea: Caracas and Beyond*, pp. 343–49. Edited by F. Christy, T. Clingan, J. Gamble, H. Knight and E. Miles. Cambridge, Mass.: Ballinger Publishing Company, 1975.

——. *The Mineral Resources of the Sea.* Amsterdam: Elsevier, 1965.

Metals Week 45 (17 Oct. 1974).

Metals Week 45 (18 March 1975).

Mikesell, Raymond F.; Bartsch, William H.; Behrman, Jack N.; Church, Phillip E.; Edwards, Gertrude G.; Gomez, Henry; Harris, William G.; Mamalakis, Markos; Wells, Donald A., Wionczek, Miguel S.; and Zinser, James E. *Foreign Investment in the Petroleum and Mineral Industries: Case Studies of Investor-Host Country Relations.* Baltimore and London: The Johns Hopkins Press for Resources for the Future, Inc., 1971.

"Mobile Units." *Offshore* 34 (July 1974):99–114.

Moncrieff, A., and Smale-Adams K. *The Economics of First Generation Manganese Nodule Operations.* Philadelphia: American Mining Congress, October 1974.

Musgrave, Richard. *The Theory of Public Finance.* New York: McGraw-Hill Book Company, 1959.

Myint, Hla. "The 'Classical Theory' of International Trade and Underdeveloped Countries." *The Economic Journal* 68 (June 1958):317–37.

Myrdal, Gunnar. *Economic Theory and Under-developed Regions.* London: Gerald Duckworth & Co., Ltd., 1957.

National Petroleum Council, Committee on Ocean Petroleum Resources. *Ocean Petroleum Resources: An Interim Report of the National Petroleum Council.* [Washington, D.C.], July 4, 1974.

National Petroleum Council. *Petroleum Resources under the Ocean Floor.* Washington, D.C., March 1969.

Naya, S. "Fluctuations in Export Earnings and Economic Patterns of Asian Countries." *Economic Development and Cultural Change* 21 (July 1973): 629–41.

"North Sea Costs and Returns." *The Petroleum Economist* 42 (December 1975):458–60.

Nove, A. *The Soviet Economy.* New York: Praeger Publishers, 1961.

"Off Borneo: SEDCO 445 Drills with Dynamic Positioning." *World Oil* 176 (February 1, 1973):31.

Offshore: The Journal of Ocean Business 30–34 (Jan. 1970–Dec. 1974).

Oil and Gas Journal 70–72 (Jan. 1972–Dec. 1974).

"Oil Industry in Poor Nations May Get World Bank Loans." *Washington Post*, 15 March 1976, p. D9.

Organization for Economic Cooperation and Development. *Development Co-operation: Efforts and Policies of the Members of the Development Advisory Committee.* [Paris], 1975.

Osgood, R., ed. *Perspectives on Ocean Policy: Conference on Conflict and Order in Ocean Relations.* Prepared for the National Science Foundation by The Johns Hopkins University, Washington, D.C., 1974.

Osgood, R.; Hollick A.; Pearson, C.; and Orr, J. *Toward A National Ocean Policy, 1976 and Beyond.* Washington, D.C.: National Science Foundation, 1975.

Papanek, G.F. "The Effect of Aid and Other Resource Transfers on Savings and Growth in LDCs." *The Economic Journal* 82 (September 1972):934–50.

Petroleum Intelligence Weekly, 14 January 1974.

Pontecorvo, G. "Reflections on the Economics of the Common Heritage of Mankind." *Ocean Development and International Law Journal* 2 (Fall 1974): 203–16.

Prebisch, Raul. "Commercial Policy in the Underdeveloped Countries." *American Economic Review Papers and Proceedings of the Seventy-first Annual Meeting of the American Economic Association* 49 (May 1959):251–73.

Resources for the Future. *Foreign Investment in the Petroleum and Mineral Industries: Case Studies of Investor-Host Country Relations.* Baltimore: Johns Hopkins University Press, 1971.

Reynolds, C. "Domestic Consequences of Export Disability. *American Economic Review* 53 (July 1963):93–102.

Roels, O.A. "The Environmental Impact of Manganese Nodule Mining." *Marine Technology Society Journal* 8 (Sept. 1974).

Salant, Stephen W. "Exhaustible Resources and Industrial Structure: A Nash-Cournot Approach to the World Oil Market." *Journal of Political Economy* (forthcoming).

Shonfield, Andrew. *The Attack on World Poverty.* London: Chatto and Windus, 1960.

Singer, Hans. "The Distribution of Gains between Investing and Borrowing Countries." *American Economic Review: Papers and Proceedings of the Sixty-Second Annual Meeting of the American Economic Association* 40 (May 1950): 473–85.

Solow, Robert M. "The Economics of Resources and the Resources of Economics." *American Economic Review: Papers and Proceedings of the Eight-sixth Annual Meeting of the American Economic Association* 64 (May 1974):1–14.

"Subsea Production System Will Be Used in 1975." *Offshore* 34 (May 5, 1974):44–45.

Sweeney, R.; Tollison, R.; and Willett, T. "Market Failure, The Common-pool Problem, and Ocean Resource Exploitation." *Journal of Law and Economics* 17 (April 1974): 179–92.

Takeuchi, K. "Unpublished Report on Copper Market Outlook." International Bank for Reconstruction and Development, Washington, D.C., 1971.

Takeuchi, K., and Varon, B. "Developing Countries and Non-Fuel Minerals." *Foreign Affairs* 52 (April 1974):497–510.

Tanzer, Michael. *The Political Economy of International Oil and the Underdeveloped Countries.* Boston: Beacon Press, 1969.

Tobin, J. "Liquidity Preference as Behavior Towards Risk." *Review of Economic Studies* 25 (February 1958):65–86.

Trilateral Commission. *A New Regime For The Oceans.* Paris, 1976.

"Unique Platform Designed for 650-Foot Water Depth." *World Oil* 176 (February 15, 1973):133.

United Nations. *Draft United Nations Convention on the International Seabed Area.* United States Working Paper, August 3, 1970.

United Nations, General Assembly, Committee on the Peaceful Uses of the Sea-Bed and the Ocean Floor beyond the Limits of National Jurisdiction. *Economic Significance in Terms of Sea-Bed Mineral Resources of the Various Limits Proposed for National Jurisdiction: Report of the Secretary General* (A/AC. 138/87), 4 June 1973.

——. *Possible Methods and Criteria for the Sharing by the International Community of Proceeds and Other Benefits Derived from the Exploitation of the Resources of the Area beyond the Limits of National Jurisdiction: Report of the Secretary General* (A/AC. 138/38), 15 June 1971.

——. *Possible Impact of Sea-Bed Mineral Production in the Area beyond National Jurisdiction on World Markets, with Special Reference to the Problems of Developing Countries: A Preliminary Assessment, Report of the Secretary General* (A/AC. 138/36), 28 May 1971.

United Nations Conference on Trade and Development. *The Effects of Possible Exploitation of the Seabed on the Earnings of Developing Countries from Copper Exports* (TD/B/484) May 28, 1974.

——. *The Effects of Production of Manganese From the Seabed, With Particular Reference to Effects on Developing Country Producers of Manganese Ore* (TD/B/483) April 23, 1974.

——. *The Effects of Production of Manganese From the Seabed, with Particular Reference to Effects on Developing Country Producers of Manganese Ore* (TD/B/483/ADD. 1) April 23, 1974.

——. *Exploitation of the Mineral Resources of the Seabed Beyond National Jurisdiction: Issues of International Commodity Policy—A Case Study of Cobalt* (TD/B/499/ADD. 1) June 26, 1973.

——. *An Integrated Programme for Commodities* (TD/B/C.1/185) June 6, 1975.

——. *Manganese Ore: Problems of Trade Liberalization and Pricing Policy* (TB/B/C.1/Cons. 7/L.2) February 1974.

United Nations, Third Conference on the Law of the Sea. *Annex to the Law of the Sea Convention: Conditions of Exploration and Exploitation* (A/Conf. 62/C1/L8) August 16, 1974.

——. *Appendix: Rules and Regulations Regarding the Conditions of Exploration and Exploitation* (A/Conf. 62/C1/L9) August 19, 1974.

——. *Basic Conditions* (A/Conf. 62/C1/L7) August 16, 1976.

——. *Economic Implications of Seabed Mineral Development in the International Area: Report of the Secretary General* (A/Conf. 62/25) May 22, 1974.

——. *Economic Implications of Seabed Mineral Development in the International Area: Report of the Secretary General* (A/Conf. 62/25) May 22, 1974.

——. *Draft Appendix to the Law of the Sea Treaty Concerning Mineral Resource Development in the International Seabed Area* (A/Conf. 62/C1/L6) August 16, 1974.

——. *Revised Single Negotiating Text.* New York Session, 15 March–7 May 1976. Reproduced by Office of Law of the Sea Negotiations, Department of State, Washington, D.C., 10 May 1976.

U.S. Atomic Energy Commission. Oak Ridge National Laboratory. *World Energy Conference: Survey of Energy Resources, 1974.* Edited by William G. Peck. New York: United States National Committee of the World Energy Conference, 1974.

U.S. Congress, Senate, Subcommittee on Minerals, Materials, and Fuels, Committee on Interior and Insular Affairs. *Current Developments in Deep Seabed Mining.* 94th Cong., 1st sess., November 7, 1975.

———. *Mineral Resources of the Deep Seabed*, pts. I and II. 93d Cong., 1st and 2nd sesss., May 1973 and March 1974.

U.S. Department of Commerce. *Economic Analysis of the Copper Industry.* Prepared by Charles River Associates, Inc. Mimeo No. PB 189–927. Washington, D.C., March 1970.

U.S. Department of Interior. *World Subsea Mineral Resources*, by V.E. McKelvey and F.F.H. Wang. Washington, D.C.: Government Printing Office, 1969.

U.S. Department of Interior. Bureau of Mines. *Minerals Yearbook.* Washington, D.C., 1973.

———. *The People's Republic of China: A New Industrial Power with a Strong Mineral Base.* Washington, D.C.: Government Printing Office [1975].

U.S. Department of Interior, United States Geological Survey. *Summary of 1972 Oil and Gas Statistics for Onshore and Offshore Areas of 151 Countries*, by Sherwood E. Frezon. Geological Survey Professional Paper 885. Washington, D.C., 1974.

———. *Summary Petroleum and Selected Mineral Statistics for 120 Countries, Including Offshore Areas*, by John P. Albers, M. Devereux Carter, Allen L. Clark, Anny B. Coury, and Stanley P. Schweinfurth. Geological Survey Professional Paper 817. Washington, D.C., 1973.

———. *The Worldwide Search for Petroleum Offshore: A Status Report for the Quarter Century 1947–72*, by Henry L. Berryhill, Jr. Geological Survey Circular 694. Washington, D.C., 1974.

U.S. Department of State. "The Third United Nations Conference on the Law of the Sea: US Delegation Report." Caracas, Venezuela, 20 June–29 August [1975]. (Mimeographed.)

U.S. Department of State. "United States Draft Environmental Impact Statement." Washington, D.C., undated. (Mimeographed).

U.S. Department of State, Agency for International Development. *Export Earnings Fluctuations and Economic Development: An Analysis of Compensatory Financing Schemes*, by Thomas Morrison and Lorenzo Perez. A.I.D. Discussion Paper No. 32. Washington, D.C.: Agency for International Development, September 1975.

———. "Pollution in the Marine Environment: Interests of Developing Countries," by W. Garling and D. Leipziger. Washington, D.C., April 1975 (Mimeographed).

U.S. Department of State. Bureau of Intelligence and Research. Office of the Geographer. *Limits in the Seas No. 36: National Claims to Maritime Jurisdictions.* 3d rev. Washington, D.C., 23 December 1975.

U.S. Department of State. Office of the Geographer. "Major Topographic Divisions of the Continental Margins" (Map). Washington, D.C., July 1970.

U.S. Department of Treasury Discussion Paper. "An Economic Analysis of the Law of the Sea Negotiations," by D. Lague and R.J. Sweeny. Washington, D.C., undated.

U.S. Department of Treasury. "Revenue Sharing With Respect to Minerals Recovered From The Seabed," by James Johnston. Washington, D.C., December 5, 1974.

U.S. "Draft United Nations Convention on the International Seabed Area: Working Paper." Washington, D.C., 30 August 1970.

U.S. Federal Energy Administration, International Energy Affairs. "The Relationship of Oil Companies and Foreign Governments." Washington, D.C., 15 February 1975. (Mimeographed.)

U.S. Federal Trade Commission. *Quarterly Financial Report For Manufacturing, Mining, and Trade Corporations.* Washington, D.C., 1975.

Verschure, P.J.M. "How Long Will Offshore Drilling Be American-Dominated?" *Oil and Gas Journal* 71 (December 31, 1973):136–41.

Wagner, Fred J. "North American Drilling Activity in 1974." *Bulletin of the American Association of Petroleum Geologists* 59 (August 1975):1270–1310.

Weiss, L. "Quantitative Studies in Industrial Organization." In *Frontiers of Quantitative Economics.* Edited by M. Intrilligator. New York: North Holland, 1971.

White, Irvin L.; Kash, Don E.; Chartock, Michael A.; Devine, Michael D.; and Leonard, R. Leon. *North Sea Oil and Gas: Implications for Future United States Development.* Norman: University of Oklahoma Press for the Council on Environmental Quality, 1973.

World Oil 176–179 (Jan. 1973–Dec. 1974).

"Worldwide Statistics." *Offshore* 30 (20 June 1970):50–57.

"Worldwide Statistics." *Offshore* 34 (20 June 1974): 88–89.

Index

About the Authors

Danny M. Leipziger is an international economist with the U.S. Agency for International Development in Washington, D.C., specializing in international finance and investment issues. He is a graduate of The City College of New York and received a doctorate from Brown University. Dr. Leipziger has conducted research on economic policy under flexible exchange rates, on international monetary reform and central bank asset behavior, and on the transfer of technology to developing countries. His papers have been published in the *Canadian Journal of Economics*, *The Oxford Bulletin of Economics and Statistics*, *World Development*, and the *Journal of Development Studies.* He recently edited a conference volume, *The International Monetary System and Developing Nations.* Dr. Leipziger has participated in Law of the Sea matters since 1973 and attended the Caracas and New York sessions of the conference.

James L. Mudge is an international economist with the U.S. Agency for International Development in Washington, D.C., specializing in international trade and payments issues. He holds a bachelor's degree in international affairs and a doctorate in economics from the University of Colorado. Dr. Mudge has conducted field research in Indonesia on capital utilization under the auspices of the University of Wisconsin Indonesia Project and has conducted policy-oriented research on the impact of increased energy costs on developing economies. He is also involved in policy and research aspects of developing country trade and employment, intermediate technology, and international petroleum economics. Dr. Mudge has been involved in issues related to seabed hydrocarbons since 1973.